ISACA Certified in Risk and Information Systems Control
(CRISC®) Exam Guide

A primer on GRC and an exam guide for the most recent and rigorous IT risk certification

Shobhit Mehta

BIRMINGHAM—MUMBAI

ISACA Certified in Risk and Information Systems Control (CRISC®) Exam Guide

Group Product Manager: Pavan Ramchandani

Publishing Product Manager: Prachi Sawant

Senior Editor: Shruti Menon

Technical Editor: Irfa Ansari

Copy Editor: Safis Editing

Project Manager: Prajakta Naik

Proofreader: Safis Editing

Indexer: Hemangini Bari

Production Designer: Vijay Kamble

Marketing Coordinator: Marylou De Mello

First published: August 2023

Production reference: 1100823

Published by Packt Publishing Ltd.

Grosvenor House

11 St Paul's Square

Birmingham

B3 1RB, UK.

ISBN 978-1-80323-690-2

www.packtpub.com

To my father, Madan, my mother, Asha, and my brother, Pinkesh, for prioritizing my education above everything else and instilling the value of discipline in me at an early age.

To my dearest wife, Dimple, for being a constant source of inspiration and my guiding light since the day I met her, and our four-legged companion, Audrey, for staying by my side on those late-night writing sprees.

Foreword

In my role as a security leader for the last five years, my top priority has been to protect my organization from cyber threats and attacks. To do this, I need to present cyber threats as key business risks to senior management and the board in a simple and effective way. Hence, it is essential to understand cyber risks and learn the art of communicating the risks that could manifest to get funding and support for the cybersecurity program.

Unfortunately, cyber risk management is underrepresented in the cybersecurity community. If you read about cybersecurity on *Medium* and other blogging platforms, you will find very detailed and helpful articles on the technical side of the field. On the other hand, there are very few blogs of the same caliber on **Governance, Risk, and Compliance** (**GRC**) and risk management. As a result, few professionals aspire to work in this domain. Shobhit aims to help simplify this field with this book. It serves as a comprehensive guide to both the principles and practicalities of GRC and the **Certified in Risk and Information Systems Control** (**CRISC**) certification.

I have been a regular reader of his blog *GRC Musings* for a long time (one of the only blogs to come under the category I mentioned before) and he brings the same clarity and thought to this book. He has distilled his learnings to help security practitioners better understand risk management and enable them to better support their organizations and upskill themselves by explaining how one can study and prepare for the CRISC certification. These topics will be especially important for aspiring GRC and risk management professionals, helping them to become well-rounded practitioners.

For anyone who wishes to be an authority on risk management, CRISC is a must. While there is ISACA's reference manual, this book fills the gap by making the subject interesting and engaging. Shobhit simplifies the jargon and helps readers understand risk management with relevant examples.

While in my first CISO role, I overheard a senior CXO equate cybersecurity to rocket science. He advised his peers to leave it to the subject matter experts. Cybersecurity today is a business risk, and to get the required support, we need to abstract the technical details and communicate why we need it. CISOs have to focus on understanding the business and the associated risks. They have to identify the potential threats and threat actors and understand the regulatory and legal landscape. Most of this comes under the GRC side of a security program. Having an effective GRC department, regardless of size, will help a CISO hit the ground running.

I would recommend this book to both aspiring GRC and risk management professionals. This will help them get a well-rounded overview of the entire GRC and risk management area. For those on the technical side, the book is a good introduction to risk management. It will help them develop as versatile security professionals and future cyber leaders.

Besides his expertise in cybersecurity, Shobhit is also an avid long-distance runner. He brings the same discipline to this book as in running – it is well-researched, meticulous, and an engrossing read. For those who love running to be fit, it is always enjoyable to go on a long run despite the pain and hard work. Shobhit's book provides a similar experience that will enrich your cybersecurity knowledge and help you grow in your career.

I am sure this book will help you in demystifying cybersecurity for management and the board while earning an important certification milestone in your cybersecurity career. Happy reading!

Vikas Yadav

CISO, Flipkart

Contributors

About the author

Shobhit Mehta is the Security and Compliance Director at Headspace, an on-demand mental health company in San Francisco, CA. Previously, he worked in different facets of security and assurance with HSBC, Deutsche Bank, Credit Suisse, PayPal, and Fidelity Investments.

He also works with ISACA to develop exam questions for CISA, CISM, and CGEIT, served as the technical reviewer for the CGEIT and CISA review manuals, and is a published author for the COBIT 5 journal.

He completed his MS in cybersecurity at Northeastern University, Boston, and holds CRISC, CISM, CISA, CGEIT, CISSP, and CCSP certifications. In his spare time, he likes to explore the inclined trails of the Bay Area, complete ultramarathons, blog on *GRCMusings*, and present at industry conferences.

I want to thank Satish Joshi and Fahad Burney, who were patient with me for four years and taught me the importance of continuous learning and attention to detail.

I also want to thank Danielle Currier and Walid Sleiman, who believed in me more than I believed in myself.

I am grateful to have met and worked for Puneet Thapliyal and Ken Stineman. Without their constant support and appreciation, this book would not have seen the light of day.

Lastly, it goes without saying that I'm extremely grateful to the professional team at Packt, including Neha, Khushboo, Shagun, Nihar, Shruti, Prajakta, Prachi, Neil, and many others, without whom I wouldn't have dared to start this book.

About the reviewer

Senjoy Joseph Panavelil is currently a Manager in the technology assurance audit practice with extensive experience in leading integrated financial statement audits and SOC 1/2 attestation engagements, performing gap analysis and SWIFT and FedLine assessments, and providing strategic recommendations for financial and banking services, retail markets, and technology industries. With a Master of Science graduate degree in information systems from the University of Florida, Senjoy is recognized for his expertise in identifying and mitigating IT risks for numerous clients, as well as his dynamic leadership skills and his ability to build trusted relationships. He enjoys using his analytical skills to work through challenging situations and come up with creative solutions.

Table of Contents

Preface   xvii

Part 1: Governance, Risk, and Compliance and CRISC

1

Governance, Risk, and Compliance   3

Governance, risk, and compliance	4	Implementing GRC using COBIT	9
What is GRC?	4	COBIT and ITIL	11
Simplified relationship between GRC components	5	A primer on cybersecurity domains and the NIST CSF	12
Key ingredients of a successful GRC program	6	Importance of IT risk management	15
GRC for cybersecurity professionals	7	Summary	16
Cybersecurity and information assurance	7		
Importance of GRC for cybersecurity professionals	8		

2

CRISC Practice Areas and the ISACA Mindset   17

CRISC exam outline	18	The ISACA mindset	24
CRISC job practice areas	19	Additional material	27
CRISC exam structure	22	Summary	28
CRISC certification requirements	23		

Part 2: Organizational Governance, Three Lines of Defense, and Ethical Risk Management

3

Organizational Governance, Policies, and Risk Management   31

IT governance and risk	32	Policy documentation	39
Key risk terminologies	32	Essential policies	41
The role of risk practitioners in IT governance	33	Exception management	42
IT risk management	33	Organizational asset	42
IT risk strategy	35	Asset valuation	44
Risk management and business objectives	36	Summary	44
Organizational structure	36	Review questions	44
RACI	37	Answers	45
Organizational culture	38		

4

The Three Lines of Defense and Cybersecurity   47

The 3LoD model	48	Risk appetite and business objectives	54
Responsibilities of 3LoD	49	Risk acceptance	54
3LoD and cybersecurity	50	Summary	55
Critical concepts for risk assessment and management	52	Review questions	55
The risk profile	52	Answers	56
Risk appetite, tolerance, and capacity	52		
Risk tolerance versus risk capacity	53		

5

Legal Requirements and the Ethics of Risk Management   57

Major laws for IT risk management   58   ISACA Code of Professional Ethics   61

Ethics and risk management   60   Summary   62

Relationship between ethics and culture   60   Review questions   62

How do ethics affect IT risk?   61   Answer   63

Part 3: IT Risk Assessment, Threat Management, and Risk Analysis

6

Risk Management Life Cycle   67

Comparing risk and IT risk   67   Correlating events and incidents   74

IT risk management life cycle   68   Summary   74

Requirements of risk assessment   71   Review questions   74

Issues, events, incidents, and breaches   73   Answers   76

7

Threat, Vulnerability, and Risk   77

Threat, vulnerability, and risk   77   Vulnerability analysis   83

The relationship between threats,   Tools for identifying vulnerabilities   84

vulnerabilities, and risk   78   Vulnerability management program   85

Understanding threat modeling   80   Summary   85

Threat modeling methods   81   Review questions   85

The importance of threat modeling   83   Answers   86

8

Risk Assessment Concepts, Standards, and Frameworks   87

Risk assessment approaches	87	Importance of a risk register	94
Which is the best approach?	89	Summary	94
Risk assessment methodologies	89	Review questions	95
Risk assessment frameworks	91	Answers	96
Risk assessment techniques	92		

9

Business Impact Analysis, and Inherent and Residual Risk   97

Differentiating between BIA and risk assessment	97	Summary	101
		Review questions	101
Key concepts related to BIA	98	Answers	102
Understanding types of risk	99		

Part 4: Risk Response, Reporting, Monitoring, and Ownership

10

Risk Response and Control Ownership   105

Risk response and monitoring	105	Summary	110
Risk owners and control owners	107	Review questions	110
Risk response strategies	108	Answers	111
Risk optimization	109		

11

Third-Party Risk Management   113

The need for TPRM	113	Managing issues, findings, and exceptions	117
Managing third-party risks	114	Summary	118
Upstream and downstream		Review questions	119
third parties	116	Answers	120
Responding to anomalies	117		

12

Control Design and Implementation   121

Control categories	121	Post-implementation reviews	125
The relationship between control categories	122	Control testing and evaluation	126
Control design and selection	123	Summary	127
Control implementation	123	Review questions	127
Control implementation techniques	124	Answers	129

13

Log Aggregation, Risk and Control Monitoring, and Reporting   131

Log aggregation and analysis	131	Risk and control reporting	135
Log sources	132	Key indicators	137
Log aggregation	133	Selecting key indicators	137
SIEM	133	Summary	138
Risk and control monitoring	133	Review questions	138
Types of control assessments	134	Answers	139

Part 5: Information Technology, Security, and Privacy

14

Enterprise Architecture and Information Technology   143

Enterprise architecture	144	Virtual private networks	152
The CMM framework	145	Cloud computing	152
Computer networks	146	Cloud computing service models	152
Networking devices	148	Cloud computing deployment models	153
Firewalls	149	Security considerations of cloud computing	153
Intrusion detection and prevention systems	149	Summary	154
The Domain Name System	150	Review questions	155
Wireless networks	151	Answers	157

15

Enterprise Resiliency and Data Life Cycle Management   159

Enterprise resiliency	159	Data life cycle management	163
Business continuity and disaster recovery	160	Comparing data management and data governance	164
Relationship between resiliency and the BCP	161	Summary	165
Recovery objectives	161	Review questions	165
Data classification and labeling	162	Answers	167

16

The System Development Life Cycle and Emerging Technologies   169

Introducing the SDLC	169	System accreditation and certification	172
Phases of the SDLC	170		
Project risk and SDLC risk	171	Emerging technologies	172

Bring your own device (BYOD)   172
Internet of Things   173
Artificial intelligence   173
Blockchain   173
Quantum computing   174

Summary   174

Review questions   174

Answers   176

17

Information Security and Privacy Principles   177

Fundamentals of
information security   178

Access management   179

Encryption   181
Types of encryption   181

Hashing   183

Digital signatures   184

Certificates   186

Public key infrastructure   188

Security awareness training   189

Principles of data privacy   189

Comparing data security
and data privacy   191

Summary   191

Review questions   192

Answers   195

Part 6: Practice Quizzes

18

Practice Quiz – Part 1   199

19

Practice Quiz – Part 2   237

Index   277

Other Books You May Enjoy   290

Preface

Welcome to this comprehensive guide to **Certified in Risk and Information Systems Control (CRISC)** by ISACA, the globally recognized authority on **Information Technology (IT)** governance and security. As organizations continue to rely more on technology to achieve their business objectives, it's becoming increasingly important for IT professionals to have the skills and knowledge necessary to manage risks effectively. The CRISC certification is designed to help IT professionals develop the expertise needed to identify, evaluate, and mitigate risks related to information systems. The certification is highly valued by employers and is considered a prerequisite for many senior-level positions.

In addition to the professional benefits of earning the CRISC certification, certified professionals have demonstrated that they possess the skills and knowledge necessary to manage information system risks effectively. This knowledge and expertise can help them make more informed decisions and improve their job performance. Furthermore, CRISC-certified professionals are in high demand and can expect to earn a higher salary than their non-certified peers. According to a survey conducted by ISACA, CRISC is the #4 top-paying certification worldwide.

This book is designed to help you achieve the CRISC certification and prepare you for the challenges of managing risks within organizations. The book is divided into three sections to provide a complete and thorough understanding of the CRISC certification and its syllabus:

- The first section provides a primer on **Governance, Risk, and Compliance (GRC)**, CRISC practice areas, and the ISACA mindset, which is essential for the certification

- The second section covers the core content of the CRISC syllabus

- The final section includes a practice quiz with detailed explanations

Whether you are a seasoned IT professional or just starting your career in IT, this book will provide you with the necessary tools and knowledge to pass the CRISC certification exam. We hope that this book will help you achieve your professional goals, improve your job performance, and take your career to the next level.

Who this book is for

This book is for professionals who are interested in obtaining the CRISC certification. The book provides a comprehensive guide to the CRISC certification and its syllabus, covering all four domains of the certification. The book is meant for professionals with differing levels of experience, from beginners to advanced practitioners.

This book is particularly relevant to professionals working in the areas of information security, risk management, and governance. It's also beneficial for individuals who are responsible for managing risks related to information systems, including IT auditors, IT consultants, and IT managers. The CRISC certification requires a minimum of three years of relevant work experience, with at least one year of experience in two or more of the four CRISC domains. Therefore, this book is recommended for professionals with some level of experience in information systems and risk management.

This book is also helpful for professionals seeking to advance their careers in the IT industry. The CRISC certification is highly valued by employers and is considered a prerequisite for many senior-level positions. By earning the CRISC certification, professionals can demonstrate their expertise in managing information system risks and increase their job prospects. It's a valuable resource that can help you achieve your professional goals, improve your job performance, and take your career to the next level.

What this book covers

Chapter 1, Governance, Risk, and Compliance, provides an introduction to GRC. This chapter includes all the lessons I learned later in my career but should have learned when I started.

Chapter 2, CRISC Practice Areas and the ISACA Mindset, provides a detailed description of the CRISC exam and practice areas. This chapter also includes my experience of attempting CRISC exams and understanding the *ISACA mindset* from both sides – as a candidate for the exam and also when I write questions for the official ISACA exam.

Chapter 3, Organizational Governance, Policies, and Risk Management, provides an introduction to organizational governance, strategy, structure, and culture. Governance is often confused with management, which is not true. This chapter continues from the lessons of *Chapter 1*.

Chapter 4, The Three Lines of Defense and Cybersecurity, provides an introduction to the concept of the **three lines of defense** and more importantly how you could draw the teachings from this model to develop your own cybersecurity program.

Chapter 5, Legal Requirements and the Ethics of Risk Management, provides an overview of major laws and regulations affecting IT risk. We will also learn about the importance of professional ethics in risk management and how it influences organizational culture.

Chapter 6, Risk Management Life Cycle, provides an introduction to the concept of risk, where you will learn how is it different from IT risk; take a deeper dive into the risk management life cycle; understand the requirements of risk assessments; learn the difference between issues, events, incidents, and breaches; and ultimately learn about how events and incidents are correlated. We will also learn how to choose different sets of controls (detective/corrective/preventive) to influence the inherent risk and optimize the residual risk.

Chapter 7, Threat, Vulnerability, and Risk, provides an introduction to the concepts of threat, vulnerability, and risk, helping you understand the relationships between each and teaching you about threat modeling and the threat landscape. We will also learn about vulnerability and control analysis, as well as vulnerability sources, and briefly touch on building a vulnerability management program.

Chapter 8, Risk Assessment Concepts, Standards, and Frameworks, builds on the knowledge from *Chapter 7*. We will learn about maintaining an effective risk register and how we can leverage already available industry risk catalogs to baseline the risk assessment program for an organization.

Chapter 9, Business Impact Analysis, and Inherent and Residual Risk, details the differences between **Business Impact Analysis (BIA)** and **risk assessments**. You will learn concepts related to BIA and the differences between inherent and residual risk, and finally, review how BIA can be used for business continuity and disaster recovery planning.

Chapter 10, Risk Response and Control Ownership, introduces the concept of risk response and monitoring and risk and control ownership, and details the risk response strategies – mitigate/accept/transfer/avoid.

Chapter 11, Third-Party Risk Management, introduces the concepts of third-party risk management and how to perform an effective third-party risk evaluation. We will also learn about issues, findings, exceptions, and how to manage them effectively.

Chapter 12, Control Design and Implementation, introduces the different types of controls, standards, frameworks, and methodologies for control design and selection and how to implement them effectively. We will also learn about several control techniques and methods to evaluate them effectively.

Chapter 13, Log Aggregation, Risk and Control Monitoring, and Reporting, provides a summary of the different methods of log sources, aggregation, and analysis. We will also learn about risk and control monitoring and reporting, and how to present them effectively.

Chapter 14, Enterprise Architecture and Information Technology, introduces the concept of enterprise architecture, the Capability Maturity Model, and IT operations, such as management and other network and technology concepts.

Chapter 15, Enterprise Resiliency and Data Life Cycle Management, provides a deep dive into the concepts of enterprise resiliency while building the foundations of a resilient architecture and data life cycle management.

Chapter 16, The System Development Life Cycle and Emerging Technologies, provides an understanding of the components of the software development life cycle and builds a foundational understanding of emerging technologies and the related security implications.

Chapter 17, Information Security and Privacy Principles, provides an understanding of information security and privacy principles, which secure the system and build trust with the users.

Chapter 18, Practice Quiz – Part 1, contains 100 review questions with a detailed explanation of each written from my experience of working with ISACA for many years.

Chapter 19, Practice Quiz – Part 2, contains additional 100 questions to solidify your understanding and ultimately set you up for success!

To get the most out of this book

To get the most out of this book, I recommend that you start with the primer section of the book, which covers the fundamentals of GRC, CRISC practice areas, and the ISACA mindset. Familiarity with industry standards and frameworks, such as **Control Objectives for Information and Related Technologies (COBIT)**, ISO 27001, and the **National Institute of Standards and Technology (NIST)** Cybersecurity Framework, is also beneficial, but not required. Additionally, we recommend that you review the CRISC certification exam syllabus before diving into the core content of the book. This will help you understand the exam objectives and the topics that will be covered in the certification exam.

As you work through the book, we encourage you to take notes, complete the review exercises at the end of each chapter, and refer back to the relevant sections when necessary. I also recommend that you take the practice quizzes at the end of the book to test your knowledge and pay equal attention to the explanation for correct and incorrect answers. By following these recommendations, you will be able to maximize your learning experience and effectively prepare for the CRISC certification exam.

Conventions used

There are a number of text conventions used throughout this book.

`Code in text`: Indicates code words in text, database table names, folder names, filenames, file extensions, pathnames, dummy URLs, user input, and Twitter handles. Here is an example: "You can find the IP address of any website by using the `ping` command."

Bold: Indicates a new term, an important word, or words that you see onscreen. Here is an example: "**Risk management** is the process of optimizing organizational risk to acceptable levels, identifying potential risk and its associated impacts, and prioritizing the mitigation based on the impact of risk on business objectives."

> **Tips or important notes**
> Appear like this.

Get in touch

Feedback from our readers is always welcome.

General feedback: If you have questions about any aspect of this book, email us at `customercare@packtpub.com` and mention the book title in the subject of your message.

Errata: Although we have taken every care to ensure the accuracy of our content, mistakes do happen. If you have found a mistake in this book, we would be grateful if you would report this to us. Please visit `www.packtpub.com/support/errata` and fill in the form.

Piracy: If you come across any illegal copies of our works in any form on the internet, we would be grateful if you would provide us with the location address or website name. Please contact us at `copyright@packt.com` with a link to the material.

If you are interested in becoming an author: If there is a topic that you have expertise in and you are interested in either writing or contributing to a book, please visit `authors.packtpub.com`.

Share Your Thoughts

Once you've read *ISACA Certified in Risk and Information Systems Control (CRISC®) Exam Guide*, we'd love to hear your thoughts! Scan the QR code below to go straight to the Amazon review page for this book and share your feedback.

https://packt.link/r/1803236906

Your review is important to us and the tech community and will help us make sure we're delivering excellent quality content.

Download a free PDF copy of this book

Thanks for purchasing this book!

Do you like to read on the go but are unable to carry your print books everywhere?

Is your eBook purchase not compatible with the device of your choice?

Don't worry, now with every Packt book you get a DRM-free PDF version of that book at no cost.

Read anywhere, any place, on any device. Search, copy, and paste code from your favorite technical books directly into your application.

The perks don't stop there, you can get exclusive access to discounts, newsletters, and great free content in your inbox daily.

Follow these simple steps to get the benefits:

1.   Scan the QR code or visit the link below:

https://packt.link/free-ebook/9781803236902

2.   Submit your proof of purchase

3.   That's it! We'll send your free PDF and other benefits to your email directly

Part 1: Governance, Risk, and Compliance and CRISC

In this part, you will get an overview of governance, risk, and compliance and how it fits into the wider gamut of information security. You will learn about the importance of IT risk management. In addition, you will learn about the CRISC exam practice areas, the types of questions you could expect, and the mindset required for ISACA certification exams.

This part has the following chapters:

- *Chapter 1, Governance, Risk, and Compliance*
- *Chapter 2, CRISC Practice Areas and the ISACA Mindset*

1
Governance, Risk, and Compliance

Dear reader, I have been in your place, thinking about which certification I should go for first. Should I begin with CISM? It seems to be the most widely recognized. Alternatively, should I consider CISA? However, I am not an auditor, so is it really necessary for me? What about CISSP? It seems rather challenging for someone trying to get certified for the first time. Finally, what about CRISC? It doesn't appear to be the most relevant for the job responsibilities in the expanding realm of IT risk management.

Congratulations! Now that you have decided on the CRISC, you have taken the most important step of deciding on your certification and are embarking on the first stage of the journey of your career growth. However, what about the study material? Should I buy the official review manual? It appears to be very dull. Should I explore technical forums or communities for more advice and hacks? Alternatively, should I conduct a search using the hashtag CRISC (#CRISC) to see if there's a one-stop blog with all the resources needed to pass the exam in one convenient location?

As I look back on all this certification preparation and reference material, I realize that the majority of them missed a key point – what is the practical application of the knowledge I will acquire as I read this book and attain the certification? If I zoom out a little, why is **governance**, **risk**, and **compliance** (**GRC**) required in an organization when the sole aim of cybersecurity is to prevent companies from attackers? Also, what is GRC in the first place?

This chapter aims to answer all these questions so that when you pass your CRISC with flying colors and boast about your certification, you don't have to worry about recalling the basic concepts of GRC and have a solid foundation of IT risk management.

In this chapter, we will cover the following topics:

- Governance, risk, and compliance
- GRC for cybersecurity professionals
- Importance of GRC for cybersecurity professionals

- A primer on cybersecurity domains and the **National Institute of Standards and Technology (NIST) Cybersecurity Framework (CSF)**.

- Importance of IT risk management

> **Important note**
>
> The content of this chapter is not directly related to the exam syllabus, but it is important to understand the concepts of GRC before learning about IT risk management and its implementation for the CRISC exam.
>
> The hope is that this chapter will provide you with enough understanding that you can differentiate between all domains of cybersecurity and can continue your journey well beyond the CRISC certification.

Governance, risk, and compliance

In this section, we'll look at the concepts of GRC, their interrelationships, and how to differentiate among them.

What is GRC?

GRC is an acronym that stands for **governance, risk, and compliance**. It can be defined as a common set of practices and processes, supported by a risk-aware culture and enabling technologies that improve decision-making and performance through an integrated view of how well an organization manages its unique set of risks.

A GRC program aims to provide organizations with an overarching framework that can be used to streamline different organizational functions, such as legal, IT, human resources, security, compliance, privacy, and more so that they can all collaborate under a common framework and set of principles instead of running individual functions and programs.

Governance is the organizational framework that helps the stakeholder set the tone for the stakeholders on the direction and alignment with business objectives. These are the rules that run the organization, including policies, standards, and procedures that set the direction and control of the organization's activities. These stakeholders can be a board of directors in large companies or senior executives in small and medium enterprises.

Risk or **risk management** is the process of optimizing organizational risk to acceptable levels, identifying potential risk and its associated impacts, and prioritizing the mitigation based on the impact of risk on business objectives. The purpose of risk management is to analyze and control the risks that can deflect an organization from achieving its strategic objectives.

Qualitative risk is defined as *likelihood * probability of impact*, whereas the **Factor Analysis of Information Risk (FAIR)** methodology is widely used for quantitative risk assessment in matured organizations.

Compliance requirements for an organization ensure that all obligations including but not limited to regulatory factors, contractual requirements, federal and state laws, certification requirements such as ISO 27001 or SOC 2 audit, and more are adhered to and any gaps in compliance are logged, monitored, and corrected within a reasonable timeframe. The entire organization must follow a standard set of policies and standards to achieve these objectives.

An integrated approach to GRC that is communicated to the entire organization ensures that the main strategies, processes, and resources are aligned according to the organization's risk appetite. A strong compliance program with the sponsorship of a senior leadership team is better suited to align its internal and external compliance requirements, leading to increased efficiency and effectiveness.

In the next section, we'll learn about the relationship between these concepts.

Simplified relationship between GRC components

I would not blame you if you found the preceding concepts tedious and confusing. It took me a good 5 years to make sense of all the concepts. The following paragraph should serve as an adage for the preceding concepts:

Governance is guidance from stakeholders (board of directors or senior leadership) to put the processes and practices in place to optimize (not reduce) the risk and comply with external and internal compliance obligations.

The following figure shows a simplistic view of GRC. It should be noted that the activities included under each pillar are not holistic and an organization may have an overlap between these activities. You should also be mindful that these activities are not standalone programs but need inputs from other pillars to be implemented successfully:

Figure 1.1 – Relationship between the components of GRC

Now that we know what GRC entails, we'll learn about the importance of various factors for a successful GRC program in the next section.

Key ingredients of a successful GRC program

A successful GRC program requires collaboration across all layers of the organization. Three major components are a must-have for successful implementation:

- **Sponsorship**: A successful GRC implementation should have the sponsorship of a senior leader such as a **Chief Information Security Officer (CISO)**, **Chief Risk Officer (CRO)**, **Chief Information Officer (CIO)**, **Chief Financial Officer (CFO)**, **Chief Executive Officer (CEO)**, or someone else. These sponsors have a wider overview of not only the organization's risk but also the industry peers across multiple verticals. Sponsorship from leadership is also important to have a risk-focused culture.

- **Stewardship**: The GRC program requires participation from all businesses and functions of an organization. Stewards play an important role in the GRC program and make information sharing across the organization easier for developing a common understanding across the organization and making relevant information available for everyone. These stewards, while translating the requirements from governance, are better able to target and address business risks. Stewards of the program are better suited to create business-oriented, process-based workflows to identify risks across functions, analyze for cascading risks, and treat them accordingly.

- **Monitoring and reporting**: It is easy to roll out a GRC program across the organization, but over time, it becomes extremely difficult to keep pace with internal and external regulations without continuously monitoring risks and controls without efficient risk indicators. It is important to enable continuous monitoring of risks and controls by using automated risk indicators and keep the stakeholders abreast of risk in business terms through business-focused indicators and reports periodically circulated to the appropriate audience with actionable insights.

  An important pillar of the monitoring function is to monitor the security controls of critical vendors and perform an ongoing assessment for each department and functional group across the enterprise to provide a holistic real-time view of risk.

In the next section, we'll learn about how to differentiate between governance and management.

Governance is not management

Those new to the GRC domain often confuse governance with management and think both are the same; however, in the realm of GRC, governance and management serve very different functions.

Governance provides oversight and is highly focused on risk optimization for the stakeholders. Governance always focuses on the following aspects:

- Is the organization doing the right things?

- Are these things done in the right away?

- Is the team getting things done on time and within budget?

- Are we continuously optimizing the risk and getting benefits?

Once these questions have been answered, the management team focuses on planning, building, executing, and monitoring to ensure that that all projects, processes, and activities are aligned with the direction and business objectives set by governance. It is expected that as management progresses in achieving these goals, the results are shared with governance (board of directors) periodically and additional inputs are taken into consideration.

GRC for cybersecurity professionals

In this section, we'll learn about cybersecurity, information assurance, and the difference between these two concepts.

Cybersecurity and information assurance

For non-cybersecurity professionals, the word cybersecurity is synonymous with hacking, but in reality, this could not be further from the truth.

There are various ways to look at cybersecurity from an outsider's view. In the industry, this is often categorized as a red team (attackers), blue team (defenders), and purple team (a combination of the red team and blue team focusing on collaboration and information sharing). For this book, I will take a different approach that is more aligned with the objectives of this book and your understanding when you prepare for the certification.

Firstly, let's segregate cybersecurity into two major practices: cybersecurity and information assurance.

In the cybersecurity realm, we consider activities such as penetration testing, vulnerability assessments, network monitoring, malware analysis, and all the other practices that require robust technical understanding and knowledge to prevent unauthorized access and disruption to the business.

The second practice, information assurance, is going to be the focus of this book. Information assurance includes activities such as policy and procedure development, risk assessments and management, data analysis, IT audits, compliance with regulatory frameworks, incident management, vulnerability management, vendor management, KPI and KRI reporting and dashboards, and all the other sub-domains that do not require extensive technical understanding. However, these practices do require thorough collaboration across all teams and a strong understanding of the fundamentals of cybersecurity concepts. These activities are important for complying with multiple federal and state regulations as well as to ensure the implementation of compliance with industry-standard practices.

Many organizations tend to completely segregate the cybersecurity and information assurance functions into different verticals altogether, where the communication between different teams and opportunities to collaborate are limited. This leads to security being seen as a gatekeeper and not an enabler.

As security is continuing to shift left – that is, being prioritized more and more in the initial stages of software development and project viability – this distinction is fading and teams using modern security tools collaborate a lot more than just meeting once a month.

As you continue with this book, you will realize that though the CRISC exam covers all concepts of cybersecurity and information assurance, the focus will primarily be on the latter as the entire purpose of the CRISC exam is to help you prepare for the IT risk management of an organization, regardless of its size.

So far, we have learned about GRC, the importance of GRC, and how to differentiate between different verticals of cybersecurity. In the next section, we'll learn about the importance of GRC for cybersecurity professionals and industry-standard frameworks to implement a GRC program.

Importance of GRC for cybersecurity professionals

As mentioned earlier, the lack of an effective GRC program makes it difficult to collaborate across all teams. An effective GRC program is the prerequisite to an effective cybersecurity program.

With the continuously increasing emphasis on privacy in the form of GDPR, CCPA, HIPAA, LGPD, and other state, national, and international regulations, the cybersecurity and information assurance teams can't work in silos. Compliance with these laws and regulatory requirements requires commitment and tenacity from all functions of the organization.

The following table shows the importance of implementing an overarching GRC framework for an organization in detail:

Non-GRC	Effective GRC
Lack of effective oversight	Effective oversight across all departments
Focus on achieving results only	Achieving results with integrity and ethics
Organizational and functional silos	Integrated decision-making
Lack of visibility	Shared technology, services, and vocabulary
Disjointed strategy	Integrated strategy

Non-GRC	Effective GRC
Duplication of efforts	Create-once, use-multiple
High costs	Optimized costs
Inefficient efforts	Efficient efforts
Lack of integrity	Culture of integrity
Wasted information	Shared and common knowledge
Fragmented information	Continuous flow of information

Table 1.1 – Importance of a GRC framework

In the next section, we'll learn about how we can use ISACA COBIT to implement a GRC program and its relationship with ITIL.

Implementing GRC using COBIT

Now that we have a good understanding of GRC and what it entails, it's important to understand how to translate this knowledge into practice.

ISACA, the certification body of CRISC, also provides a comprehensive framework called **Control Objectives for Information and Related Technology (COBIT)** to bridge the gap between governance, technical requirements, business objectives and risks, and control requirements.

The latest version of COBIT (COBIT 2019) guidance from ISACA focuses on providing elaborate guidance on managing risk, optimizing resources, and creating value by streamlining all business objectives.

There are four publications under the COBIT 2019 framework:

- **Introduction and Methodology**: This is the fundamental document for implementing the COBIT framework that details governance principles, provides key concepts and examples, and lays out the structure of the overall framework, including the COBIT Core Model.

- **Governance and Management Objectives**: This publication contains a detailed description of the COBIT Core Model and its 40 governance and management objectives. These are then defined and matched with the relevant processes, enterprise goals, and governance and management practices.

- **Design Guide: Designing an Information and Technology Governance Solution**: This publication provides essential guidance on how to put COBIT to practical use while offering perspectives for designing a tailored governance system for an organization.

- **Implementation Guide: Implementing and Optimizing an Information and Technology Governance Solution**: This document, combined with the COBIT 2019 Design Guide, provides a practical approach to specific governance requirements.

COBIT Core includes 40 governance and management objectives that have defined purposes that are mapped to specific core processes. These objectives are primarily divided into five categories:

- **Evaluate, Direct, and Monitor (EDM)**: EDM has five objectives that focus on a few specific, governance-related, areas. These include alignment of enterprise and IT strategies, optimization of costs and efficiency, and stakeholder sponsorship.

- **Align, Plan, and Organize (APO)**: APO's 14 objectives include managing organizational structure and strategy, budgeting and costs, the HR aspect of IT, vendors, **service-level agreements (SLAs)**, risk optimization, and data management.

- **Build, Acquire, and Implement (BAI)**: The 11 BAI objectives are focused on managing changes to data and assets while ensuring end user availability and capacity needs are met.

- **Deliver, Service, and Support (DSS)**: DSS contains six objectives and mostly aligns with the IT domains. DSS is focused on managing operations, problems, incidents, continuity, process controls, and security.

- **Monitor, Evaluate, and Assess (MEA)**: MEA has four objectives related to creating a monitoring function that ensures compliance for APO, BAI, and DSS. These objectives include managing performance and conformance, internal control, external requirements, and assurance. Notably, MEA differs from EDM by concentrating on the monitoring function from an operational standpoint, whereas EDM monitors from a governance (or top-down) approach.

The following figure shows the five domains and 40 COBIT Core processes:

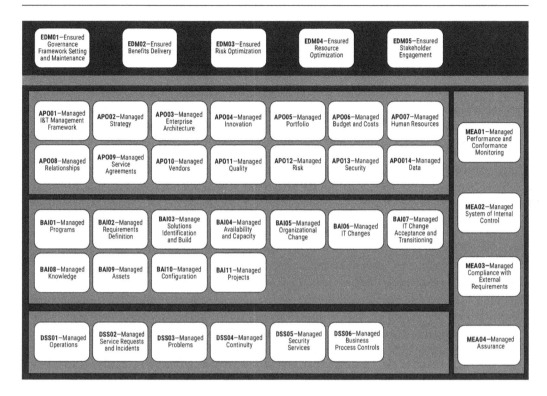

Figure 1.2 – COBIT 2019 Core Model (COBIT® 2019 Framework: Governance and Management Objectives ©2019 ISACA. All rights reserved. Used with permission.)

> **Important note**
>
> Detailed guidance on ISACA introduction and methodology is available at no cost to members and non-members on the ISACA website: `https://www.isaca.org/resources/cobit`.

COBIT and ITIL

This section would not be complete without understanding the relationship between COBIT and ITIL.

ITIL is a framework designed to standardize the selection, planning, delivery, and maintenance of IT services within an enterprise. The goal is to improve efficiency and achieve predictable service delivery.

ITIL and COBIT are both governance frameworks but serve different purposes. ITIL primarily aims to fulfil service management objectives, whereas COBIT is globally recognized for both enterprise governance and IT management.

On their own, each framework is extremely successful in offering custom governance while delivering quality service management. When paired together, however, COBIT and ITIL have the potential to dramatically increase value for customers as well as internal and external stakeholders.

The COBIT framework helps identify *what* IT should be doing to generate the most value for a business, ITIL prescribes *how* it should be done to maximize resource utilization within the IT purview. Even though the frameworks are different, they do have multiple touchpoints – for example, from the COBIT domain, BAI, process BAI06 Managed IT Changes is equivalent to ITIL Change Management; process BAI10 Managed Configuration is equivalent to ITIL Configuration Management, and so on.

A major differentiation between COBIT and ITIL is that COBIT covers the entire enterprise, ensuring that governance is achieved, stakeholder value is ensured, and holistic approaches to governing and managing IT are accomplished, whereas ITIL is focused entirely on IT service management. COBIT aims to achieve its objectives through policies, processes, people, information, and culture and organizational structures, services, and applications that are implemented and integrated under a single overarching framework for ease of integration and customization, whereas ITIL provides prescriptive guidance on implementing these objectives.

In the previous section, we learned about the importance of ISACA COBIT for implementing a GRC program and its relationship with ITIL. In the next section, we will learn about multiple cybersecurity domains and the NIST CSF.

A primer on cybersecurity domains and the NIST CSF

There are many, many ways to think about cybersecurity domains and this could very well be a book in itself. The purpose of this section is to provide an overview of common cybersecurity domains and what they entail.

For the sake of simplicity and aligning it with a common industry standard, this section is aligned with the NIST CSF.

The NIST CSF divides the cybersecurity domain into five main categories, namely, Identify, Protect, Detect, Respond, and Recover:

- **Identify**: There is an old saying in the cybersecurity world – *You cannot protect what you do not know exists*. The Identify category of the CSF emphasizes developing the organization's understanding to manage cybersecurity risk to systems, assets (including people), data, and the capabilities to do so.

This activity is important for prioritizing the organization's efforts and resources in consistency with its overall risk management strategy and business goals. This function stresses the importance of understanding the business context, the resources that support critical functions, and the related cybersecurity risks. The activities in Identify include the following:

- Identification of physical, software, and people assets to establish the basis of an asset management program

- Identification of established cybersecurity policies to define the governance program, as well as identifying legal and regulatory requirements regarding the cybersecurity capabilities of the organization

- Identification of the organization's business environment and critical systems, including the role of critical vendors in the supply chain

- Identification of asset vulnerabilities, threats to internal and external organizational resources, and risk response activities to assess risk

- Implementation of a risk management strategy, including identifying risk appetite and tolerance

- Identification of vendor risk management strategy, including priorities, constraints, risk tolerances, and assumptions used to support risk decisions associated with managing supply chain risks

- **Protect**: Once the assets and critical processes have been identified, the appropriate safeguards (controls) must be developed and implemented to ensure the delivery of critical infrastructure services. This function is dedicated to identifying controls that outline appropriate safeguards to ensure the delivery of critical infrastructure services and supports the ability to limit or contain the impact of a potential cybersecurity event. The activities in Protect can be seen here:

  - Perform security awareness training for all staff and additional role-based and privileged user training.

  - Implement protections for identity management and access control within the organization, including physical and remote access. In the case of an external data center or using cloud services, implement robust controls such as complex passwords, the use of VPNs, and multi-factor authentication.

  - Establish data security protection consistent with the organization's risk strategy and criticality of assets to protect the confidentiality, integrity, and availability of information.

  - Implement processes and procedures to maintain and manage the protection of information systems and assets.

  - Protect organizational resources through maintenance, including remote maintenance activities.

  - Manage technology to ensure the security and resilience of systems, consistent with organizational policies, procedures, and agreements.

- **Detect**: Proactively detecting and deterring potential cybersecurity incidents is critical to a robust information security program. This function defines the appropriate activities to proactively identify the occurrence of a cybersecurity event and involve the relevant teams as soon as the threat vectors are identified. The activities in Detect can be seen in the following list:

  - Detect anomalies across all system events and act on them before they cause substantial harm to the assets

  - Implement tools for continuous monitoring and detection (also known as the Security Operations Centre) to monitor critical events, tune the systems to reduce false positives, and gauge the effectiveness of protective measures, including network and physical activities

- **Respond**: Once an event has indeed materialized and caused the incident, the organization should be prepared to contain and respond using manual as well as automated processes. This function aims to develop such systems, train the staff on incident response, and ensure that incidents can be resolved within the agreed timeframe and with minimum disruption to the system. The activities in Respond include the following:

  - Manage communications with internal and external stakeholders during and after an event

  - Analyze the incident to ensure effective response and supporting recovery activities including forensic analysis and determining the impact of incidents

  - Ensure incident response planning processes are agreed upon with relevant staff, executed at the time of the incident, and lessons learned are improved to prevent the incident in the future

  - Perform mitigation activities to prevent the expansion of an event and to resolve the incident

  - Implement improvements by incorporating lessons learned from such responses and ensure the staff is trained on the new practices

- **Recover**: This function identifies appropriate activities to renew and maintain plans for resilience and to restore any capabilities or services that were impaired due to a cybersecurity incident. The activities in Recover can be seen here:

  - Ensure that the organization has a recovery plan process in place that is tested within an acceptable time frame and that procedures to restore systems and/or assets affected by cybersecurity incidents are in place

  - Implement the lessons learned while responding to incidents and review those with relevant stakeholders

  - Internal and external communications are coordinated during and following the recovery from a cybersecurity incident, and new areas of risk are continuously added and acted upon

The following figure summarizes the NIST CSF functions:

Identify

- Identify assets and processes that need protection

Protect

- Implement appropriate controls to ensure the protection of assets

Detect

- Implement appropriate mechanisms to identify the occurrence of cybersecurity incidents

Respond

- Develop techniques to contain and minimize the impact of cybersecurity events/incidents

Recover

- Implement the appropriate processes to restore capabilities and services within an agreed time frame

Figure 1.3 – Simplified NIST CSF functions

Each of these domains is further segregated into multiple subdomains that are outside the scope of this book. I highly encourage you to familiarize yourself with the NIST CSF subdomains and their relationship with COBIT.

> **Important note**
>
> COBIT has custom frameworks for several specific use cases, including a framework for implementing the NIST CSF. A set of such publications can be found on the ISACA website at https://www.isaca.org/resources/cobit.

Importance of IT risk management

Now that we've discussed a fair bit about GRC, the domains of cybersecurity, and the NIST CSF, it is important to understand the implications of IT risk management for an organization.

In an enterprise risk management function, there can be a myriad of risks such as strategic risk, environmental risk, market risk, credit risk, operational risk, compliance risk, reputational risk, and more.

All the preceding risks can be impacted by IT risks in three major ways:

- **IT value enablement risk**: The delivered projects did not create the expected value, leading to a loss of shareholder value and opportunities that could have materialized

- **IT program and project delivery risk**: Projects are not ready to be delivered as agreed with the internal and external stakeholders, leading to inconsistency with the overall strategy

- **IT operations and service delivery risk**: Delivered services are not in compliance with the SLAs agreed upon at the inception of the project

All the preceding impacts have cascading effects on other areas of the organization. An overarching governance framework implementation can prevent these risks from materializing.

Summary

At the beginning of this chapter, we learned that governance is the guidance from stakeholders (board of directors or senior leadership) to put the processes and practices in place to optimize (not eliminate) the risk and comply with external and internal compliance obligations. Then, we looked at the key ingredients of a successful GRC program, including sponsorship, stewardship, monitoring, and reporting. We concluded this chapter by understanding the ISACA COBIT framework for a GRC program implementation and its relationship with ITIL and providing a primer on cybersecurity domains and the NIST CSF. Now, you should be well equipped to start conversations regarding a GRC program implementation and speak about its value with the senior leaders in your organization.

In the next chapter, we will switch gears and learn about the CRISC practice areas and the ISACA mindset to answer the CRISC exam questions.

2
CRISC Practice Areas and the ISACA Mindset

If the previous chapter was all about learning about governance, risk, and compliance, and why they are required, this chapter will focus on preparing you for the main goal of this book – to pass the ISACA **Certified in Risk and Information Systems Control** (**CRISC**) exam.

The CRISC certification aims to advance your career by helping you understand the impact of IT risk and how it relates to your organization. The CRISC certification demonstrates the holder's ability to identify and evaluate IT risk, propose strategies to mitigate risk optimally, and help the enterprise accomplish its business objectives.

The ISACA website (`https://www.isaca.org/credentialing/crisc`) provides an apt description of the certification: *The CRISC certification validates your experience in building a well-defined, agile risk-management program, based on best practices to identify, analyze, evaluate, assess, prioritize, and respond to risks. This enhances benefits realization and delivers optimal value to stakeholders.*

Since its inception in 2010, more than 30,000 professionals worldwide have earned the CRISC credential. The CRISC credential enables an IT risk manager to showcase their competence and ability to design, implement, monitor, and maintain effective risk-based information systems controls.

In addition to the preceding attributes, the CRISC certification also does the following:

- Proves your skills and knowledge in using governance best practices and continuous risk monitoring and reporting

- Enhances business resilience and stakeholder value and allows you to gain increased credibility with peers, stakeholders, and regulators

- Ensures you are recognized as a professional with the skills and experience to provide value and insight from an overall organizational perspective on both IT risk and control

- Establishes a common language to communicate within IT and to other stakeholders (privacy, legal, people operations, human resources, and more) throughout enterprises about risk

- Ensures you will understand information system control design, implementation, control monitoring, and maintenance
- Affirms your ability to plan and implement appropriate control measures and frameworks that further mitigate enterprise risk without impeding innovation
- Ensures your ability to provide guidance to technical and management teams alike on the prioritization of resources needed to achieve business objectives

This chapter aims to assist you with developing a mindset so that you think of all the preceding points when you encounter a decision-making scenario in your organization and also when you attempt scenario-based questions in the exam.

In this chapter, we will cover the following topics:

- CRISC exam outline
- CRISC job practice areas
- CRISC exam structure
- CRISC certification requirements
- The ISACA mindset
- Additional material

With that, let us move on to the CRISC exam outline.

CRISC exam outline

ISACA is well known for updating the job practice areas; for instance, they modify the syllabus for exams of all flagship certifications at least every five years or update the existing practices to meet the professional demands of the industry. The current CRISC exam outline was updated in August 2021 to keep up with the industry requirements for IT risk management jobs.

The following table shows a breakdown of the old and new job practice areas:

Domains	Old CRISC job practice	New CRISC job practice
Domain 1	IT Risk Identification (27%)	Governance (26%)*
Domain 2	IT Risk Assessment (28%)	IT Risk Assessment (20%)
Domain 3	Risk Response and Mitigation (23%)	Risk Response and Reporting (32%)*
Domain 4	Risk and Control Monitoring and Reporting (22%)	Information Technology and Security (22%)*

* indicates revised domains

Table 2.1 – Old and new job practice areas

As per the revised CRISC syllabus, the following figure shows a breakdown of domains in terms of the number of questions and percentage:

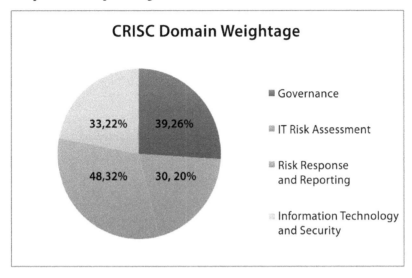

Figure 2.1 – CRISC domains

As you can see in the previous practice areas, the job practices were only highly focused on risk management (identification, assessment, response, mitigation, and control monitoring). The revised categories are broadly divided into governance, IT risk assessment, response and reporting, and general IT and security.

If you recall the previous chapter, this should make a lot more sense in terms of why an entire chapter on governance, risk, and compliance was required even before starting to prepare for the CRISC exam.

In the next section, we'll take a dep-dive into the practice areas of the revised outline.

CRISC job practice areas

I have combined the job practice areas to ease the flow of reading and the logical structure of the book. The following list summarizes the practice areas and their corresponding chapters in this book:

Domain 1 – Governance

- **Organizational governance**:
  - *Chapter 3, Organizational Governance, Policies, and Risk Management*:
    - Organizational strategy, goals, and objectives
    - Organizational structure, roles, and responsibilities

- Organizational culture

- Policies and standards

- Business processes

- Organizational assets

- **Risk governance**:

  - *Chapter 4, The Three Lines of Defense and Cybersecurity:*

    - Enterprise risk management and the risk management framework

    - Three lines of defense

    - Risk profile

    - Risk appetite, tolerance, and capacity

  - *Chapter 5, Legal Requirements and the Ethics of Risk Management:*

    - Legal, regulatory, and contractual requirements

    - Professional ethics of risk management

Domain 2 – IT risk assessment

- **IT risk identification**:

  - *Chapter 6, Risk Management Life Cycle:*

    - Risk events (e.g., contributing conditions and loss result)

  - *Chapter 7, Threat, Vulnerability, and Risk:*

    - Threat modeling and threat landscape

    - Vulnerability and control deficiency analysis (e.g., root cause analysis)

    - Risk scenario development

- **IT risk, analysis, evaluation, and assessment**:

  - *Chapter 8, Risk Assessment Concepts, Standards, and Frameworks:*

    - Risk assessment concepts, standards, and frameworks

    - Risk register

    - Risk analysis methodologies

- *Chapter 9, Business Impact Analysis, and Inherent and Residual Risk:*

  - Business impact analysis

  - Inherent, residual, and current risk

Domain 3 – Risk response and reporting

- **Risk response**:

  - *Chapter 10, Risk Response and Control Ownership:*

    - Risk treatment/risk response options

    - Risk and control ownership

  - *Chapter 11, Third-Party Risk Management:*

    - Third-party risk management

    - Issue, finding, and exception management

    - Management of emerging risk

- **Control design and implementation**:

  - *Chapter 12, Control Design and Implementation:*

    - Control types, standards, and frameworks

    - Control design, selection, and analysis

    - Control implementation

    - Control testing and effectiveness evaluation

- **Risk monitoring and reporting**:

  - *Chapter 13, Log Aggregation, Risk and Control Monitoring, and Reporting:*

    - Risk treatment plans

    - Control types, standards, and frameworks

    - Risk and control monitoring techniques

    - Risk and control reporting techniques (heatmap, scorecards, and dashboards)

    - **Key Performance Indicators (KPIs)**

    - **Key Risk Indicators (KRIs)**

    - **Key Control Indicators (KCIs)**

Domain 4 – Information technology and security

- **Information technology principles**:

  - *Chapter 14, Enterprise Architecture and Information Technology:*

    - Enterprise architecture

    - IT operations management (e.g., change management, IT assets, problems, and incidents)

    - Project management

  - *Chapter 15, Enterprise Resiliency and Data Life Cycle Management:*

    - **Disaster Recovery Management** (**DRM**)

    - Data life cycle management

  - *Chapter 16, The System Development Life Cycle and Emerging Technologies:*

    - The **System Development Life Cycle** (**SDLC**)

    - Emerging technologies

- **Information security principles**:

  - *Chapter 17, Information Security and Privacy Principles:*

    - Information security concepts, frameworks, and standards

    - Information security awareness training

    - Business continuity management

    - Data privacy and data protection principles

In the next section, we will learn about the CRISC exam structure.

CRISC exam structure

All ISACA exams, including the CRISC, consist of 150 questions covering all the areas discussed in the previous section. At the time of publication, the four domains have different weightages for exam questions.

Here is a summary of the number of questions per domain:

CRISC job practice	Weightage	Questions
Governance	26%	39
IT risk assessment	20%	30
Risk response and reporting	32%	48
Information technology and security	22%	33

Table 2.2 – Number of questions across domains

You will have 4 hours to answer 150 questions, which, in my experience, is ample time to answer all the questions. The exam passing score is 450 on a scale of 200 to 800.

> **Important note**
>
> The CRISC exam does *not* have any negative scoring, that is, there are no deductions for incorrect answers. The best strategy to attempt the questions that you are unsure about is to mark them for later review but answer them before submitting the exam.

You will receive your results as soon as the exam is submitted. The detailed scores are sent to your registered email within 7 business days. At that time, you should be able to log in to ISACA and look at the detailed scoring per domain.

In the next section, we will learn about the CRISC certification requirements.

CRISC certification requirements

Passing the exam with a minimum of 450 marks is one of the requirements for attaining the CRISC certification.

Once you pass the exam, the next step is to submit a CRISC application form, which should be endorsed by a colleague, previous manager, or someone who knows you in a professional capacity, attesting that you have a minimum of 3 years of cumulative work experience in at least two of the four domains. The work experience must be gained within the 10-year period preceding the application date for certification. After passing the exam, you have 5 years to apply for the certification.

Once the certification is issued, you will also have to maintain a minimum of 120 **Continuing Professional Education (CPE)** hours over a 3-year cycle and report an annual minimum of 20 CPE hours to keep the certification active.

Lastly, you will have to comply with ISACA's Code of Professional Ethics to uphold the professional and personal conduct of members of the association and its certification holders.

In the next section, we will learn about the ISACA mindset and how to differentiate between and attempt the knowledge-based versus scenario-based questions in the exam.

The ISACA mindset

After attempting all the major ISACA certifications and then being actively involved on the other side of the table, that is, writing questions for the official exams, I think I have a fair understanding of the rationale for answering the ISACA questions. In the ISACA working group, and in multiple forums on the internet, you will often hear about developing the ISACA mindset before attempting the exam.

> **Important note**
>
> The ISACA mindset involves understanding the rationale behind why a certain question is asked and what would be the MOST appropriate answer. When you read the question, you should ask yourself what concept the exam is trying to test and assume the role of an IT risk manager while answering the question. Once you have a fair understanding of the reasoning for the question in the first place, you should look for the answer that looks the closest to an ideal answer. It should be noted that all four options in the ISACA exam will seem to be the right answer, but your aim should be to identify the answer that closely matches closely the real-world scenario and practicability.

We can broadly divide the ISACA exam questions into two categories – **knowledge-based** and **scenario-based**.

Knowledge-based questions rely primarily on the candidate's fundamental knowledge of the terms referenced in the question, whereas scenario-based questions rely on a candidate's ability to think through the scenario, step into the role the question is referencing, and then choose the most appropriate answer. From my experience of attempting and writing these questions, practical experience in the domains becomes extremely important, so you can step into the shoes of the role mentioned in the question and relate that to real-world scenarios.

Let us look at an example of each with explanations to understand the ISACA mindset and put it into practice.

1. **Question** – Accountability for information categorization lies the *most* with:

   A. Security analyst

   B. **Senior management**

   C. End user

   D. Custodian

Though all the options look correct, the most appropriate answer is *B. Senior management.*

There are a few keywords in the preceding question that we need to be mindful of when answering it.

The first keyword is *accountability*. As we will learn in the upcoming chapters, the senior management role is responsible for ensuring information is categorized, for instance, having appropriate data classification and labeling. This is important so that appropriate controls can be put in place as per the category of data (think highly critical data, such as company financial reports, and contracts versus public information such as marketing material). The security analyst will be responsible for enforcing the categorization but will not be accountable for it. Similarly, the end user will be responsible for adhering to the control requirements but will not be accountable. A custodian is a distractor and is not accountable for deciding the controls.

The second keyword in the question is *categorization*. Again, the senior management role is responsible for deciding the data classification and type of labeling for the organizational data. The other options are distractions.

With a quick glance at the options, we can definitely eliminate the security analyst and end user, as both of them will not be accountable for ensuring the categorization of information but will be responsible for it. The other options might be a bit confusing but by the end of the book, it should be fairly clear that the custodian is responsible for ensuring the controls are put in place and adhered to but is not accountable for designing them. And with that, we can come to the final answer for this question, which is senior management. As I mentioned at the beginning of the section, while attempting these questions, it's important to know the fundamental terms. Knowing them should make the choices fairly easy.

The following is an example of a scenario-based question that needs more thinking and understanding from a practical standpoint:

2. **Question** – A risk assessment has been completed for a major vendor of an organization. In the case of a disagreement between the business manager and the information security manager, which of the following would be the *best* approach for the information security manager to use to resolve the disagreement?

    A. Acceptance of the business manager's decision on the risk to the corporation.

    B. Acceptance of the information security manager's decision on the risk to the corporation.

    C. **Review of the risk assessment with senior management for final input and risk acceptance.**

    D. Creation of a new risk assessment to resolve the disagreement.

You should expect 50% (75) of the questions to be scenario-based like the preceding example. Some of these questions might have a more complex or simpler scenario, but those questions are designed to test your practical experience in the real world.

For this example, the correct answer is *C. Review of the risk assessment with senior management for final input and risk acceptance.*

Let us dive into each of the options and see why they are not the correct answers.

Option *A. Acceptance of the business manager's decision on the risk to the corporation* is incorrect. Firstly, the question mentions that it is a major vendor critical to the business or internal processes of the organization, and hence, the failure of that vendor could lead to high risk to the organization. Secondly, the business manager's primary job in the organization is not to assess the risk to the organization but get the job done. From their point of view, it would be a hurdle to not have the vendor onboarded as early as possible and implement the new process. With the preceding reasoning, this option can be safely eliminated.

Option *B. Acceptance of the information security manager's decision on the risk to the corporation* might look to be the correct answer in this scenario, but it is not. Opposite to the business manager's job of getting the task done and implementing the process at the earliest moment, the information security manager's job is to keep the organization secure. But as mentioned in *Chapter 1*, the job of a risk manager is not to eliminate the risk but optimize it. In this scenario, if we decide to accept the security manager's decision, the organization will not be able to use the vendor, and it may impact a major business process and, hence, the business objectives. This would not be an ideal solution for the business.

Option *D. Creation of a new risk assessment to resolve the disagreement* may again look like the correct answer, but it is not the best approach. A risk assessment has already been performed by the information security manager, and the question does not elaborate on any additional criteria that might help the security manager while performing the new risk assessment. Therefore, the new risk assessment will lead to the same results as the earlier risk assessment and an impasse.

Option *C. Review of the risk assessment with senior management for final input and risk acceptance* is the *best* answer from the available options. It is already given in the question that a risk assessment has been performed and the security manager and business manager do not agree with the results of the risk assessment. This is a major vendor for the organization, so it is critical to the business too. However, in the case of a conflict, the best-case scenario will be to put the pros and cons in front of the senior management, let them weigh in on the importance of the vendor and the identified risks, and require them to make the final decision.

In the official exam, it is important to focus on keywords, such as *best*, *most*, and *least*, to choose the most appropriate answer from all the options.

> **Important note**
>
> In a real-world scenario, there are a few additional next steps that should be taken by the information security manager:
>
> 1. Document any mitigation plans that cannot be implemented immediately but will be considered in the future to reduce the risk.
>
> 2. Document a formal sign-off from the senior management mentioning that the residual risk is acceptable for the business.
>
> 3. Identify any corrective action plans that might help reduce the overall risk for the organization and agree on a plan with the vendor to resolve those within agreed timelines.
>
> 4. Recommend available alternatives to the business manager in lieu of the proposed vendor, if any.

In the CRISC exam, you could expect similar questions to the preceding examples. It is important to do as many practice questions as possible to develop the mindset that we discussed at the beginning of this section. I have included a practice quiz at the end of each chapter to assist with the development of your thought process and two additional practice quizzes toward the end of the book.

In the next section, we will look at some additional material that will supplement the content of this book.

Additional material

This book covers all the content, tips, and practice quizzes you need to pass the ISACA CRISC exam. As discussed in the *CRISC job practice areas* section, some sections are bundled to ease the flow of the content in a way that should make the most logical sense. As much as I would want to keep this as a single source for all the knowledge you would need to pass the exam, there are two additional resources that I would highly recommend to supplement your learning:

- **ISACA CRISC Review Manual, 7th Edition**

  This is the official review manual from ISACA, the governing body that creates and conducts the exam. Almost everyone I have ever spoken to has given negative feedback on the ISACA material due to its dry nature and monotonous writing, but I would highly recommend you go through it at least once. ISACA has a way of phrasing common terminologies as well as information security concepts. The content might not be as interactive as this book, but it is important to learn the official terminologies to keep any surprises at bay during the exam. A single detailed read of the review manual in addition to this book should be sufficient for the review material.

- **CRISC Review Questions, Answers, and Explanations Manual, 6th Edition**

  This is the other official source that I would highly recommend for supplementing the practice quizzes. The questions in this book are the closest you can answer outside the exam environment. Questions are divided into subsequent domains and have detailed explanations for each correct and incorrect option.

  Though this book will have a similar flavor of questions, do keep in mind that these questions are written by a single individual (yours truly), whereas the official CRISC review questions manual is written by experts around the world who are vetted multiple times to meet the CRISC exam standard. There is a very low probability that questions from this book will appear as is in the final exam, but completing and understanding the rationale for each correct answer will definitely help you develop the correct mindset to answer the official exam questions.

Both resources are available for direct purchase from the ISACA website.

Summary

At the beginning of this chapter, we learned about the CRISC exam domains and the weightage for each domain in the exam. We learned about the official CRISC job practice areas and how they are reflected in the structure of this book. Then, we learned about the CRISC exam structure in detail and the passing criteria, that is, a minimum score of 450 and 3 years of experience in at least 2 domains to achieve the official CRISC credential and an additional requirement of maintaining the CPEs once the credential is achieved. We also noted that there is no negative marking/deduction for incorrect answers in the exam, so your best bet is to answer all the questions even if you are not 100% sure about the answer. At the end of the chapter, we learned about the ISACA mindset, the types of questions that will be asked (knowledge-based versus scenario-based), how to read and identify the keywords in the questions, and tips on how to eliminate incorrect answers. Lastly, we looked at the additional material that will supplement the preparation for the exam.

The next chapter will be the official start of the CRISC syllabus, and we will learn in detail about the importance of organizational governance, policies, and risk management. I hope you are as excited to learn as I am to support you in this journey!

Part 2: Organizational Governance, Three Lines of Defense, and Ethical Risk Management

In this part, you will get an understanding of organizational governance and how organizational culture affects risk, and learn about the importance of asset management. In addition, you will learn how the three lines of defense fit into cybersecurity, how to differentiate between risk appetite and risk tolerance, and how ethics and culture affect IT risk.

This part has the following chapters:

- *Chapter 3, Organizational Governance, Policies, and Risk Management*
- *Chapter 4, The Three Lines of Defense and Cybersecurity*
- *Chapter 5, Legal Requirements and the Ethics of Risk Management*

Organizational Governance, Policies, and Risk Management

This chapter is a detailed version of the topics we briefly touched on in *Chapter 1*. As we learned in *Chapter 1*, the purpose of an organization is to create value for the stakeholders, shareholders, and customers. This is achieved by aligning the enterprise's mission, objectives, and strategy. Similarly, organizational structures and leadership are required to establish objectives that support their mission and satisfy stakeholders and customers. The board of directors establishes the strategy and the enterprise derives its principles from this plan.

Organizational leaders support the enterprise's objectives and decision-making by evaluating the risk and benefits associated with specific investments. Senior managers identify the capabilities that contribute value to the organization's strategy.

This is then translated into policies, standards, and procedures, which are essential to putting the objectives into practice for all employees.

This chapter aims to introduce the concept of organizational governance, strategy, structure, and culture. Governance is often confused with management, which is not true. For concrete organizational governance, each organization should have an appropriate structure in place, with stakeholders, board members, management, the leadership team, stewards, and more for support that's ingrained in its culture to ensure its independence.

We will also learn about establishing policies, standards, and procedures (collectively referred to as policy documentation). We will also look at how the culture of an organization is highly influenced by its leadership and how their support to implement an appropriate culture of IT risk management is extremely important.

In this chapter, we will cover the following topics:

- IT governance and risk
- Organizational structure

- Organizational culture
- Policy documentation
- Organizational asset

With that, let's dive into IT governance and risk.

IT governance and risk

In any organization, governance establishes the requirements for meeting stakeholder needs and delivering value for the stakeholders. The purpose of having a governance structure is to have accountability for the business alignment and day-to-day operations of the organization. The purpose of **governance of enterprise IT (GEIT)** is to leverage technology to support and optimize enterprise needs. GEIT helps organizations address common pain points, such as applicable laws, regulations, and compliance, and stay abreast with the latest technologies and innovations. IT governance empowers organizations and helps establish and monitor accountability for IT activities to ensure that investments in IT are aligned with the business objectives and promote stakeholder value generation.

> **Important note**
> An organization can generate value for stakeholders by realizing the benefits of investments, risk optimization, and resource optimization.

Key risk terminologies

We briefly touched on this in *Chapter 1*, where we learned that risk is the product of likelihood and impact. Two additional key terms that are closely associated with risk need to be understood before we move on with this chapter: threats and vulnerabilities. *Threats* are defined as any circumstance or event that could potentially harm the systems and cause risk to the organization, while *vulnerabilities* are the weaknesses that threats exploit to materialize the risk. A *threat actor* is a malicious person or group that could cause harm to the organization.

Consider the example of a malware infection in an organization. In this case, the threat actor is the person who created the malware in the first place, the threat to the organization is that the laptops could get infected, and the vulnerability is the weakness of either not installing an antivirus or an employee unknowingly opening an infected file. Here, the risk is infected laptops having an impact on day-to-day operations or causing the disclosure of confidential information:

Figure 3.1 – The asset, threat, vulnerability, and risk relationship

In the next section, we will review the role of risk practitioners in IT governance.

The role of risk practitioners in IT governance

Risk practitioners play an important role in providing the correct solutions for enterprise IT. Once the organization agrees on the business requirements of IT, the risk practitioners are responsible for providing accurate and timely information to the relevant decision-makers so they can make well-informed decisions. Investments in IT often take considerable time and effort of all the involved resources and proper due diligence is required to assist major IT decisions. The decision to use certain tools and technologies may be a one-time exercise, but the risk practitioner should continuously identify, assess, mitigate, and monitor the risks that pose a major risk to the organization and impact its ability to provide services to the customer.

IT risk management

IT risk management is the practice of understanding the business goals and overall risk strategy, as well as guiding the IT strategy to align with organizational goals and priorities with minimal risk. The IT strategy needs to be supported by the available resources, technical maturity, and available budget.

Like enterprise risk management, IT risk management is a cyclical process that consists of the following steps:

Figure 3.2 – IT risk management life cycle

The following is a brief description of each step of the IT risk management life cycle:

1. **IT risk identification**: This is the first step of IT risk management and includes determining the level of risk per the enterprise's risk appetite and tolerance. It is important to document the risk identification efforts and include the major threats to organization assets, including people, processes, and technologies.

2. **IT risk assessment**: This step requires analyzing, evaluating based on existing controls, and prioritizing risks identified in the earlier phase to determine the controls required per the criticality of the assets.

3. **Risk response and mitigation**: Once the risks have been identified and assessed, it is important to respond to such risks and mitigate them accordingly. The risks can be treated by mitigating, accepting, transferring, or avoiding them.

4. **Risk and control monitoring**: After a successful treatment strategy has been agreed upon with the relevant stakeholders, it is important to continuously monitor the risks and control to ensure the changing technological factors will not lead to a change in risk status.

5. **Risk reporting**: This phase is often ignored by risk practitioners but is equally important as the other phases. Once the risk treatment cycle is complete, it is important to report the key indicators to senior management. This enables the management team to analyze the key themes and put in additional controls or reduce the scope of controls if the asset does not require that set of controls.

In the next section, we will review how the IT strategy is related to the IT risk strategy.

IT risk strategy

An IT risk strategy closely follows the IT strategy. The IT risk strategy is an essential part of enterprise risk management as it supports all the key functions and businesses of the organization. The IT risk strategy should encompass mechanisms to provide benefits to the stakeholders, program delivery, IT operations, service delivery, and cyber and information security risk. The IT risk has to be measured by its impact on IT services and business operations alike.

The following table shows a list of common IT risks that the IT risk manager should be aware of:

IT Risk	Description
Access risk	Disclosure of information to unauthorized users, leading to a loss of confidentiality.
Availability risk	Information not being available to authorized users when needed.
Cyber and information security risk	Lack of logical and/or physical controls to ensure confidentiality, integrity, availability, and privacy of data.
Emerging technology risk	Risk of negative/adverse impact of the implementation of new technology that has not been fully adopted in the industry.
Infrastructure risk	Risk of IT infrastructure not being able to support the needs of the business, customers, and users. Infrastructure includes hardware, software, network, people, processes, systems, or ad hoc resources such as vendors required to provide the services.
Integrity risk	Risk of incomplete, incorrect, or inaccurate data caused by lack of sufficient controls.
Investment or expense risk	Risk of failure of investment due to excessive spending or lack of results on investments.
Program/project risk	Risk of failure of the program or project due to a lack of commitment and accountability.
Relevance risk	Risk of unavailability of information to the right recipients at the right time to allow for actions.
Schedule risk	Risk of the project not being completed on time.
Talent risk	Risk of key resources leaving the organization.

IT Risk	Description
Third-party risk	Risk of vendors failing to provide adequate services or getting compromised.
Fourth-party risk	Risk of the vendor's vendor failing to provide adequate services or getting compromised.

Table 3.1 – Common IT risks

> **Important note**
>
> Senior management's support and sponsorship is the most critical aspect of effective risk management. Without the support of senior management, it will be extremely hard for the risk managers to obtain sufficient budget and legitimacy for the risk management program. Senior management drives the culture of an organization, and without them communicating the importance of risk management efforts, the risk management program will almost always be unsuccessful.

Risk management and business objectives

The risk manager needs to align the risk management efforts per the business objectives and not vice versa. This is critical to obtaining sponsorship from senior management as well as the support of peers, departments, and business partners so that the risk management efforts are supported by everyone across the organization. This can be achieved by having a clear understanding of the objectives of the enterprise and maintaining active communication with the leadership team. Initiatives such as a steering committee with participants from all groups help the risk manager and business alike to focus on high-risk areas that can adversely impact the business objectives.

Organizational structure

The success of an organization's risk management program depends on the sponsorship and support of the senior management. Different departments may lead risk management programs and to make risk-conscious decisions, senior management needs to combine all the individual programs in an enterprise risk program, often referred to as **enterprise risk management (ERM)**.

The IT risk manager needs to be acquainted with the ERM program and establish roles and responsibilities for all relevant stakeholders. This can often be performed by a tool called RACI.

RACI

RACI is an effective tool for determining the roles and responsibilities of a project with several stakeholders with varying priorities. There are four main roles under the RACI method:

Role	Description
Responsible	The individual or team responsible for performing the risk management efforts
Accountable	The individual who is accountable for the success of the project by providing adequate resources to the *responsible* group
Consulted	Subject matter experts or business process owners with expert knowledge of the domain
Informed	An individual or group that may or may not be a key stakeholder but will be affected by the success or failure of the efforts

Table 3.2 – Roles included in the RACI method

Let's take an example of a common risk management practice for organizations building the program. Consider that a medium-sized organization would like to conduct a NIST **Cybersecurity Framework (CSF)**-based risk assessment across the organization. Conducting this assessment would be a substantial effort and the risk manager would have to assign responsibilities to the team members for the successful and timely completion of the assessment.

In this case, the risk manager is the person *responsible* for conducting the risk assessment and may include additional team members to support them throughout the efforts. The risk manager needs the sponsorship of senior management, such as the CISO or a similar role, to ensure that they receive appropriate resources to carry out the assessment and also garner the support of other departments for which this activity might just be an additional process in their regular responsibilities. Since this assessment is related to their area (cyber assessment), the CISO will also be *accountable* for completing the project.

The risk manager may or may not have the technical capabilities to understand the entire control implementation per the NIST guidelines and will *consult* with the subject matter experts in the related areas. They will also have to reach out to individual business process owners to understand the assets under consideration and recommend appropriate controls. Lastly, the entire assessment might have come from a request from the board of directors as a step to mature the enterprise risk management program, and would be keenly interested in being *informed* about the progress of the project.

It is important to learn that though the preceding example speaks in selective terms of a risk manager or CISO, it is very well possible that the names of the roles will differ from one organization to the other. However, the organizational structure of oversight, management, and employees performing the task will remain the same:

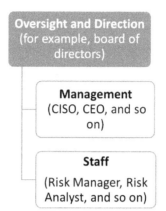

Figure 3.3 – Example of an organizational structure

In the next section, we will review the organizational culture and how it's important for organizational risk management.

Organizational culture

In all my experience so far with several organizations, the only common thing was a huge emphasis on driving a resilient risk culture.

If you take away one lesson from this chapter, let the following be it.

> **Important note**
> Nothing impacts an organization's behavior toward risk management more than its culture and nothing impacts an organization's culture more than senior management.

An organization's culture toward risk management can be divided into five parts:

- **Vulnerable**: Neither senior management nor employees care about the organization's risk and the response is always after the risk has materialized.

  For example, the IT admin only updates the antivirus after the infection has happened.

- **Reactive**: The response is based on the complaints of the employees or when required for compliance with contractual or non-contractual obligations.

  For example, the organization is undergoing an external audit and needs to install an antivirus to pass the audit.

- **Compliant**: The organization has responsibilities defined but the response is based on external compliance requirements such as HIPAA, SOX, and others.

  For example, a healthcare organization chooses to perform a HIPAA risk assessment to comply with HIPAA requirements.

- **Proactive**: The organization's senior management is well informed about the risks to the organization and supports the team by providing sufficient sponsorship and resources for effective risk management.

  For example, an organization proactively seeks an external audit validation such as ISO 27001 and aims to improve the gaps identified.

- **Resilient**: The organization has clear accountabilities established and communicated throughout and emphasizes risk management in everything they do.

  For example, the developers are fully trained in secure coding practices and always have *Security by Design* and *Privacy by Design* at the top of their minds.

The most important factor in establishing a resilient culture toward risk is the sponsorship of senior management. No amount of training or motivation will work if the emphasis on a risk-driven culture is not communicated from the top and exhibited by the leadership team.

The risk culture of an organization directly impacts the behavior of the employees. This behavior then affects the organizational policies, standards, and adherence to them.

Policy documentation

For the sake of simplicity policies, standards, procedures, and guidelines are collectively referred to as policy documentation:

- **Policies**: Policies are high-level statements of management intent from an organization's executive leadership that are designed to influence decisions and guide the organization to achieve the desired outcomes.

  Policies are enforced by standards and further implemented by procedures to establish actionable and accountable requirements.

> **Important note**
> Policies are a business decision, not a technical one. Technology determines how policies are implemented.

- **Standards**: Standards are *mandatory* requirements concerning processes, actions, and configurations that are designed to satisfy control objectives.

Standards are intended to be granular and prescriptive to establish minimum security requirements that ensure systems, applications, and processes are designed and operated to include appropriate cybersecurity and privacy protections per the organization's risk appetite and requirements.

- **Procedures**: Procedures are a documented set of steps necessary to perform a specific task or process in conformance with an applicable standard.

  Procedures help address the question of how the organization operationalizes a policy, standard, or control. Without documented procedures, there can be defendable evidence of due care practices.

  Procedures are generally the responsibility of the process owner/asset custodian to build and maintain but are expected to include stakeholder oversight to ensure applicable compliance requirements are addressed.

  Procedures are often referred to as **standard operating procedures (SOPs)**.

- **Guidelines**: Guidelines are recommended practices that are based on industry-recognized secure practices. Guidelines help augment standards when discretion is permissible.

  Unlike standards, guidelines allow users to apply discretion or leeway in their interpretation, implementation, or use.

The following diagram shows the relationship between the various parts of the policy documentation:

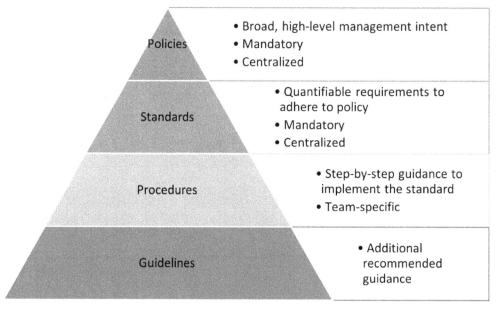

Figure 3.4 – Policy documentation hierarchy

Let's understand the relationship between the policy documentation with the example of implementing encryption.

The overall corporate policy will say that employees should encrypt all data at rest. The standard related to the policy will state the industry-acceptable encryption algorithms acceptable for encryption, such as AES-256. The procedure will provide step-by-step guidance on how to encrypt this data and lastly, the guideline will provide additional guidance, such as how to proceed in case of an error.

Essential policies

> **Important note**
> This section is not required for the CRISC exam but is essential to know for the role of an IT risk manager.

Developing essential policy documentation is the foundation of an effective risk management program. An IT risk manager should liaise with the relevant stakeholders to develop key policies. The following policies are a good list to help you start the policy documentation:

- **Information security policy** (**ISP**): This policy provides a holistic view of all the security controls for all the assets – physical or data. The ISP includes controls for unauthorized users and unauthorized access to data, programs, systems, and the organization's infrastructure. The ISP may be the most important policy as it forms the basis of all the other policies and controls to be developed.

- **Access control policy**: This policy includes requirements for authenticating users, authorizing, modifying, and removing users and access using role-based access control. This policy also guides you on how to provide privileged access and how to treat service accounts.

- **Password policy**: This policy includes requirements for minimum length and complexity, restricting the use of old passwords, and expiration. This policy also includes requirements for password storage and password requirements for privileged accounts.

- **Data classification policy**: This policy dictates how the data should be secured and what controls should be put in place to protect the data. Data classification helps determine what baseline security controls should be put in place to safeguard the data.

- **Physical security policy**: This policy defines the requirements for protecting information and technology resources from physical and environmental threats to reduce the risk of loss, theft, damage, or unauthorized access to those resources.

- **Acceptable use policy**: This policy dictates how company resources should be used. This policy applies to internal employees, as well as contractors.

- **Backup policy**: This policy defines an organization's requirements for backing up company data and systems. The backup policy should dictate the extent and frequency of backups per the criticality of the data.

- **Logging and monitoring policy**: This policy documents the requirements for logging user activity and the procedures for reviewing the logs.

- **Risk assessment policy**: This policy documents the procedures for performing periodic risk assessments. The policy includes how the organization identifies potential threats (logical and physical), analyzes the significance of risks associated with the identified threats, and determines the mitigation strategies for the identified risks. The process of identifying the risks and performing control assessments is also called a *risk and control assessment*.

- **Change management policy**: This policy documents the procedures for making changes to IT infrastructure and applications. The policy includes the standard processes for requesting, testing, and approving changes before implementing them in production.

- **Incident response policy**: This policy documents the procedures that security personnel should follow when a security incident has been identified. This procedure also includes detection, containment, evaluation, and reporting.

- **Business continuity plan (BCP)**: This is a plan to continue operations if a place of business is affected by different levels of disaster. This can be localized short-term disasters, to days-long, building-wide problems, to the permanent loss of a building. It should be noted that the **disaster recovery plan (DRP)** is a part of the BCP.

Ideally, all policies should be reviewed and approved at least on an annual basis or after a major change in business process or infrastructure to ensure they still support the management's intent.

Exception management

Sometimes, employees will have to accept the risk of not adhering to the policy and deviation from the standards to achieve the business objectives. This is acceptable as the function of information security is to enable the business and not be a roadblock.

In cases such as these, exceptions should be raised by the requestor and approved explicitly by the business process owner or a responsible delegate. These exceptions are logged centrally and often reviewed annually to ensure that any exceptions to the policy that are not required can be removed.

Organizational asset

There is a saying in information security – *you can't protect what you don't know exists*. The entire purpose of risk management is to protect assets.

Assets are anything that provides value to the organization. These can be either tangible such as equipment, physical media, laptops, and so on, or intangible such as data, knowledge, reputation, people, and more.

Here is the list of the major assets of any organization:

- **People**: For any organization, their people are the greatest asset. Organizations are vulnerable to the loss of a key employee who may be the only person with the expertise and know-how in a specific area. Failure to identify key resources and effective cross-training in the absence of a key employee could lead to an ineffective transition or loss of business.

- **Technology**: Using outdated systems and technology could lead an organization into a precarious situation where they are vulnerable to malware infections and unpatched systems. It is also important to dispose of the media and delete data with industry-standard mechanisms such as NIST 800-88.

- **Data**: In modern times, data is the new currency. The first step to protecting any data is to identify the importance of the data for the organization. This can be performed by identifying and socializing data classification across the organization, such as highly confidential, confidential, internal, public, and so on. The controls on the data should be directly proportional to the importance of data for the organization.

- **Intellectual property**: Trademarks, copyrights, patents, and trade secrets are considered intangible assets for an organization. Failure to protect intellectual property could lead to a loss of competitive advantage.

The following table describes the key terms relevant to intellectual property:

Term	Definition
Trademark	A sound, color, logo, saying, or another distinctive symbol that is closely associated with a certain product or company.
Copyright	Protection of writing, recordings, or other ways of expressing an idea. The idea itself may be common, but the way it was expressed is unique, such as in a song or book.
Patent	Protection of research and ideas that led to the development of a new, unique, and useful product to prevent the unauthorized duplication of the patented item.
Trade secret	A formula, process, design, practice, or another form of secret business information that provides a competitive advantage to the organization that possesses the information.

Table 3.3 – Terms related to intellectual property

In the next section, we will learn the importance of asset management and valuation.

Asset valuation

Asset valuation directly impacts the cost of controls that are put in place to protect the asset. It is important to have appropriate data classification and data labeling to ensure that the cost of controls does not exceed the cost of assets.

The risk manager should be able to identify the importance of assets to the organization and recommend appropriate controls per the effective value of the asset. An organization can choose either a qualitative or a quantitative method to assign value to an asset. Organizations also base the value of an asset on the impact of the loss of confidentiality, integrity, or availability and its effect on the organization.

Summary

At the beginning of this chapter, we learned about the relationship between IT governance and risk. We learned about the IT strategy and how the IT risk strategy acts as a supporting mechanism for achieving business objectives. We then learned about the relationship between threats, vulnerabilities, assets, risks, and the IT risk management life cycle. Then, we learned about the importance of organizational structure and setting the tone for risk management from the top and how is it related to the organizational culture. Finally, we learned about policy documentation and the importance of asset classification and labeling for implementing appropriate controls per the asset valuation.

In the next chapter, we will look at the importance of three lines of defense in cybersecurity and why is it required to establish accountability and avoid conflicts of interest.

Review questions

1. Which of the following is the first step in IT risk management?

    A. IT risk assessment

    B. Risk response and mitigation

    C. IT risk identification

    D. Risk reporting

2. Oversight and direction are the responsibility of who?

    A. Management

    B. Senior staff

    C. IT risk manager

    D. Board of directors

3. The high-level intent of an organization's security practices is best established by which of the following?

   A. Procedures

   B. Policies

   C. Guidelines

   D. Processes

4. Which of the following groups should be reached out to for expert guidance on the working group?

   A. Responsible

   B. Accountable

   C. Consulted

   D. Informed

5. An IT risk manager recently joined an organization that had quite a few security incidents in the past. Which of the following should be the *primary* focus of the IT risk manager in the first few months of their tenure to reduce similar incidents in the future?

   A. Cite individuals who caused the incidents

   B. Understand the organization's culture toward risk

   C. Draft the strategy for IT risk management

   D. Complete all the security awareness training

Answers

1. **C**. IT risk identification is the first step of IT risk management. You cannot protect what you do not know exists.

2. **D**. Oversight and direction are the responsibility of the board of directors. The rest of the options are important for risk management, but they do not provide the required oversight.

3. **B**. Policies are the high-level intent of organizations for security practices.

4. **C**. Consulted working groups are the subject matter experts for guidance on initiatives in their domain of expertise.

5. **B**. The risk manager needs to understand the culture of the organization in the first place before drafting the strategy for risk management. Option A depends on the severity of the incident, while D is important but will not be the primary focus for reducing future incidents.

4

The Three Lines of Defense and Cybersecurity

In the wake of the financial crisis, the **Institute of Internal Auditors (IIA)** came up with a model for risk management and called it the **Three Lines of Defense (3LoD)** model. 3LoD traces its origins to the managing of operational risk in large organizations, especially financial institutions. However, in the recent past, this has slowly gained traction in the cybersecurity world, too. The main objective of the 3LoD framework is to ensure the effective segregation of duties for all business functions and better accountability for the stakeholders of those functions.

As we saw in earlier chapters, one of the major functions of IT risk management is to have an effective delineation between the risk owners (think the board of directors) and risk practitioners (think senior management).

In the 3LoD framework, the goal is to have business functions segregated based on the duties they perform for effective risk management.

In simple terms, 3LoD can be summarized as follows:

- First LoD – operational management
- Second LoD – risk monitoring and oversight
- Third LoD – internal or external audit

The aim of this chapter is to introduce the concept of 3LoD and, more importantly, how you can draw on the learnings from this model to develop your own cybersecurity program.

We will also learn about establishing an organization's risk profile, risk appetite, risk tolerance, the differences between risk appetite and risk tolerance, and their relationships with risk capacity. Additionally, we will look at how the risk appetite of an organization is highly influenced by the business objectives.

In this chapter, we will cover the following topics:

- The 3LoD model
- 3LoD and cybersecurity
- Critical concepts for risk assessment and management
- Risk tolerance versus risk capacity
- Risk appetite and business objectives

With that, let us dive into the 3LoD model.

The 3LoD model

As mentioned in the previous section, the purpose of the 3LoD model is to ensure appropriate segregation and accountability for individual business owners and other functions.

Let's take a step back for a moment to establish the ownership of the risk – if there is a risk to the business, who will be the owner of the risk? The business owner. Therefore, the business owner will also be the risk owner as per the 3LoD model. Since the business owners are responsible for the day-to-day operational management of the business, they will be considered the first LoD for any risk that might occur to their business.

Now, these business owners might know a lot about the business, the risks, and the control environment, but they might not be the experts in remediating those risks. This is where the second LoD comes in. The second LoD is the risk monitoring and oversight function. They work closely with the first LoD to ensure that these risks are mitigated with the least disruption to the business.

Once the first LoD and second LoD perform their responsibilities, there needs to be a level of assurance that there is proper segregation of roles and responsibilities in these duties, and this is provided by either the internal audit or the external audit. These internal and external auditors, thus, form the third LoD for an organization:

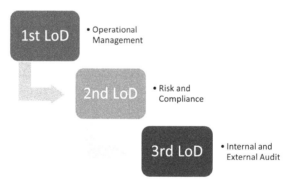

Figure 4.1 – 3LoD model

In the following section, we will look at the responsibilities of each LoD in detail.

Responsibilities of 3LoD

3LoD differs in its responsibilities. This section will go into the details of each LoD:

- The first LoD – **operational management**:

  - Business owners with a thorough understanding of the business and control environment

  - Responsible for implementing risk management practices in the respective businesses

  - Responsible for internal control monitoring and ensuring that control deficiencies are highlighted and addressed

  - Responsible for implementing corrective actions when required and ensuring that risk is always within risk tolerance

  - Ultimate risk owners

- The second LoD – **risk and compliance functions**:

  - Also known as risk monitoring and oversight

  - Primary responsibility is to support the first LoD by developing and providing guidance

  - Responsible for developing the risk management framework and policy documentation (policies, standards, and procedures) in accordance with the business objectives

  - Responsible for communicating the risk management framework across the organization and obtaining buy-in from key stakeholders

  - Monitoring the first LoD activities to ensure compliance with the organization's risk management framework is maintained at all times

  - Developing **Key Risk Indicators** (**KRIs**) and keeping the relevant stakeholders abreast of credible threats faced by the organization

- The third LoD – **audit**:

  - Includes both internal and external audits

  - The most important aspect is "independence" from the first LoD and the second LoD

  - Responsible for evaluating the effectiveness of the first and second LoD activities

  - Responsible for providing independent opinions of the organization's conformance against the internal or external risk management framework

Let us look at an example to concrete our understanding of the three LoD responsibilities.

When I worked for a large multinational bank, there were Oracle databases with thousands of orphan accounts (orphan accounts are those accounts that do not have an owner – either they left the organization and their account was not revoked, or they moved to a new department and did not need the previously provisioned account, but apparently no one took the pain to remove their previous account).

Ideally, the process should have been pretty straightforward – after someone leaves the org/team, revoke the account or disable the access, and set up automation to delete the account after a set period or keep it for business reasons such as maintaining audit trails. But of course, if all of this was implemented, we would not be learning from this example.

In this case, when the external auditor came to audit the systems, they realized that thousands of such database accounts were active with no owner in sight, and they raised it as an audit finding. This is exactly the responsibility of the third LoD – to highlight the key risks and ensure such results are noted in the audit report so that action can be taken accordingly.

In this case, the second LoD failed to perform its duties. Ideally, the termination procedures should have been monitored to ensure that such accounts were removed from the systems and any gaps in relevant KRIs should have been highlighted to the relevant stakeholders.

The first LoD, who owns the systems and the databases, were also the risk owners. They failed to perform their duties of complying with the relevant policies and procedures, leaving the accounts active even after the employees no longer needed them.

Therefore, the responsibilities of 3LoD can be summed up as follows:

Figure 4.2 – Simplified 3LoD responsibilities

In the next section, we will look at how this model can be translated and utilized for cybersecurity functions.

3LoD and cybersecurity

So far, we have looked at the 3LoD model from an overarching view of enterprise risk management. In this section, we will translate those responsibilities for cybersecurity functions.

Please be aware that an *official* 3LoD and cybersecurity model is not mainstream, but the following figure should give you a good sense of the responsibilities for each role:

1st LoD	2nd LoD	3rd LoD
• Deploy Intrusion Detection and Prevention systems	• Document policies, standards, and procedures	• Provide independent ongoing evaluations of preventive and detective measures related to cybersecurity
• Securely configure the network to manage and protect the network traffic flow	• Perform periodic user reviews	
	• Perform privileged user reviews	• Evaluate IT assets of users with privileged access for standard security configurations, malicious websites and software, and data exfiltration
• Deploy data protection and loss prevention systems with related monitoring	• Segregation of Duties analysis	
	• Termination process review	
• Restrict least privileged roles	• Classify data and design least-privilege access roles	• Track diligence of remediation
• Implement vulnerability management with internal and external scans	• Plan/test business continuity and participate in disaster recovery exercises	• Conduct cyber risk assessments of service organizations, third parties, and suppliers (1st and 2nd LoD share this ongoing responsibility)
• ...and all other **Operational** activities	• ...and all other **Monitoring and Oversight** activities	

Figure 4.3 – 3LoD and cybersecurity responsibilities

In the preceding figure, we can see that the first LoD is doing the on-ground work and owns the responsibilities of the operational management of technical activities such as conducting penetration tests and vulnerability assessments, restricting user access rights, and more.

The second LoD is focused on performing oversight and monitoring in the form of documenting the policies/standards/procedures as per the business objectives, conducting access reviews, planning and conducting the **Business Continuity Planning** (**BCP**) exercise, reviewing termination procedures, and more.

Finally, the third LoD is the internal or external auditor performing audits as per the organization's risk management framework or industry standards such as ISO 27001, AICPA Trust Principles (SOC 2), or HITRUST CSF.

> **Important note**
> Please note that the preceding list of activities in *Figure 4.3* does not encompass all the activities that exist in the realm of cybersecurity. There are many, many other operational and oversight activities that we could not cover in the scope of this chapter.

In the next section, we will learn about an organization's risk profile, appetite, tolerance, and capacity.

Critical concepts for risk assessment and management

This section is the essence of the critical concepts that will be widely used across this book, your day-to-day job, and any ISACA exam including the CRISC.

The risk profile

The purpose of the risk management function is to optimize the risk decisions for an enterprise. The risk profile is the overall risk exposure of the organization to any type of risk. There are many factors that could impact the risk profile of an organization, such as new regulations, changes in the underlying technology, changes in the business objectives, mergers and acquisitions, direct or indirect competitors, and more. This is all part of the enterprise risk profile and will impact all businesses and functions of the organization.

The IT risk profile of an organization is the overall identified **IT risk** to which the enterprise is exposed. Similar to the enterprise risk profile, the IT risk profile can be dependent on many external factors such as emerging threats, internal and external malicious actors, incidents, changes in IT regulations such as newly introduced privacy frameworks, new or acquired assets, supply chain risks, and more.

With so many factors affecting the risk profile, it is important to develop, socialize, and report actionable KRIs so that recurring and/or high risks can be identified and remediated as soon as is practical.

> Important note
> The most important aspect of effective KRIs is that they should be actionable. A set of KRIs that are reported to stakeholders for driving inferences but are not actionable will not be of much use to senior management.

Risk appetite, tolerance, and capacity

The aspects of risk appetite, tolerance, and capacity are intricately related and should be understood in detail as they are the foundations of risk assessments for an organization.

Let us start with the definition of each, and then we will learn about their relationship with business objectives:

- **Risk appetite** – this is the amount of risk an organization is willing to accept to achieve its objectives

- **Risk tolerance** – this is the acceptable level of variation that the management is willing to allow to achieve its objectives

- **Risk capacity** – this is the amount of risk an organization can afford to take without its continued existence being called into question

> **Important note**
>
> Risk tolerance is a slight deviation from the acceptable risk levels that are acceptable to achieve the objectives of the organization; however, risk appetite and tolerance should always be less than the risk capacity.

The following diagram shows the relationship between risk appetite, tolerance, and capacity:

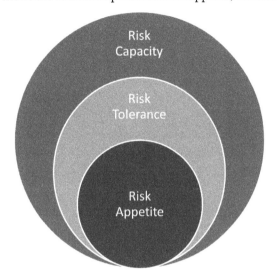

Figure 4.4 – Relationship between risk appetite, tolerance, and capacity

Now that we have a good understanding of risk appetite, capacity, and tolerance, in the next section, we will learn, in depth, about the differences between risk tolerance and risk capacity.

Risk tolerance versus risk capacity

I have seen IT risk practitioners use the phrases risk tolerance and risk capacity interchangeably, but this is not correct.

An organization going a little beyond the risk appetite is still within the risk tolerance, which is manageable as long as there are some compensating controls in place. However, when the risk tolerance crosses a certain threshold, it enters into the territory of risk capacity. As we saw in the earlier definition, anything over risk capacity could impact the existence of the organization, and that is something that has to be avoided at all costs.

That said, an organization can still operate as intended within its risk tolerance and under the risk capacity, but its existence will be in question if it crosses the risk capacity.

In the following section, we will see the relationship between risk appetite and business objectives.

Risk appetite and business objectives

The risk appetite of an organization should be agreed upon with the relevant stakeholders. It is important to align risk appetite with the objectives of the business to ensure that high-risk areas providing more value to the business are getting more resources than low-risk, low-reward processes.

The best way to align risk appetite with the business objectives is to translate it into a number of standards and policies to contain the risk level within the boundaries set by the risk appetite. With changing business conditions, these boundaries need to be regularly adjusted or confirmed.

Risk acceptance

Risk appetite and tolerance need to be defined, approved, and clearly communicated by the senior management with a process in place to review and approve any exceptions. These exceptions are formally documented in the form of risk acceptance.

As important as it might be to keep the risk levels within acceptable thresholds, there might be some edge cases where the rewards outweigh the cost of risk treatment and the business must accept the consequences if the risk materializes.

For the risk acceptance process, it is important to document the risk owner (business owner) accepting the risk, countermeasures that could help with reducing the risk in the future, the duration of risk acceptance, and a final sign-off from the executive team ensuring that all relevant teams and stakeholders seeking a risk acceptance are aligned with the risk acceptance. All these risk exceptions should be logged in a separate system of record (such as a GRC tool or spreadsheet) and should be revisited, at the very least, on an annual basis.

> **Important note**
>
> It is extremely important to note that the risk acceptance could exceed the risk appetite and tolerance but should *not* exceed the risk capacity, as that could threaten the continued existence of the organization.

Consider the following example to understand the relationship between risk appetite and risk acceptance.

During an internal risk assessment, the software development team identifies that some of the critical data for some customers managed by a third-party vendor is not encrypted at rest. This causes an uproar within the team, and the initial instinct of the IT risk manager is to classify this as a high risk. The team determines that the company could lose some of the contracts due to this lack of encryption, and it may cost $100,000 in revenues on those lost contracts. Upon further investigation, it is observed that the data stored with the third-party vendor is completely deidentified and cannot be reversed back to its original state. The total cost and resources for encrypting the deidentified data will be $500,000.

With all the numbers in place, the business owner of the contracts is willing to *accept* the risk, as the cost of remediating the risk ($500,000) far outweighs the rewards ($100,000). More so, there are compensating controls in place such as data deidentification, which limits the risk of data exposure, and hence, the risk can be accepted after a final sign-off from the business owner.

Summary

At the beginning of this chapter, we learned about the 3LoD model and the responsibilities of each LoD. Then, we reviewed how we can translate the 3LoD model for IT risk management and cybersecurity. In the next section, we switched gears to learn about the importance of the risk profile, appetite, tolerance, capacity, the relationship between all of them, and how to distinguish between risk tolerance and risk capacity. Another major area covered in this chapter was how to determine the risk appetite for a business and the process for formal risk acceptance.

In the next chapter, we will learn about the legal, regulatory, and contractual requirements, along with ethical risk management.

Review questions

1. In the 3LoD model, which LoD is responsible for risk monitoring and oversight?

    A. The first LoD

    B. The second LoD

    C. The third LoD

    D. All of the above

2. What is the primary responsibility of the third LoD?

    A. Policy and procedure development

    B. Provide independent assurance of controls

    C. Perform periodic user reviews

    D. Restrict least privileged roles

3. The amount of risk an organization is willing to accept is known as _____::

    A. Risk tolerance

    B. Risk capacity

    C. Risk profile

    D. Risk appetite

4. The information security manager has performed a risk assessment and provided recommendations for enhancing the controls of the **Business Process Owner** (**BPO**). After much deliberation, the BPO has decided to accept the risk. The BEST reason for the BPO to accept the risk is _____:

    A. Difficulty to implement the suggested controls

    B. Unavailability of resources to implement the controls

    C. Cost of control implementation outweighs the cost of assets

    D. Budgetary constraints

5. Which of the following statements is correct?

    A. Breaching risk tolerance could threaten an organization's existence

    B. Breaching risk capacity could threaten an organization's existence

    C. Risk tolerance and capacity are not related at all

    D. Risk tolerance and capacity are the same

Answers

1. **B.** The second LoD is responsible for risk monitoring and oversight. Please refer to the *Responsibilities of 3LoD* section for additional details.

2. **B.** The third LoD is primarily responsible for an internal and external audit, which is an independent assurance of controls. The keyword here is *independent*, which is the utmost requirement of a third LoD function.

3. **D.** An organization should be able to accept the risk within the risk appetite. Any risk above the risk appetite should be reduced to an acceptable level by implementing adequate controls.

4. **C.** This is the fundamental aspect of risk management to ensure that the cost of control implementation should not outweigh the cost of assets. For example, it does not make sense to put a $1,000 lock on a $500 bicycle.

5. **A.** This is the only correct statement in this question. Option **B** is incorrect as that should be risk tolerance. Option **C** is incorrect as we learned in the *Risk appetite, tolerance, and capacity* section that both are related. Option **D** is a distractor and incorrect.

5
Legal Requirements and the Ethics of Risk Management

For any organization, legal, regulatory, and contractual requirements play a major role in how the organization is governed, and this is no different from how IT risk is managed. Failure to comply with the local laws and applicable regulations could lead to severe penalties causing monetary and reputational damages. Multinational organizations need to be wary of the local, federal, and international regulations in addition to the specific laws of each industry, making it extremely difficult for the organization to comply with all the laws of each location.

Historically, these requirements were delegated to business owners to ensure controls against business processes such as financial fraud and corporate governance controls were in place. However, in the last two decades, the responsibility has also equally shifted on the IT side with the responsibilities of confidentiality/integrity/availability of data, privacy requirements for consumer data, protection of intellectual property rights, and so forth.

The aim of this chapter is to provide an overview of such laws and regulations. Additionally, we will learn about the importance of professional ethics in risk management and how it influences an organizational culture.

In this chapter, we will cover the following topics:

- Major laws for IT risk management
- Ethics and risk management
- How do ethics affect IT risk?
- ISACA Code of Professional Ethics

With that, let us dive into the major laws guiding IT risk management.

Major laws for IT risk management

Compliance is a fundamental consideration for any organization dealing with information security and privacy. Implementing and monitoring internal controls is critical for an organization that handles information that falls within the scope of many continuously evolving state, federal, and industry requirements.

IT incidents such as data leakage or ransomware could lead an organization to not only fall out of compliance but also deal with major financial and reputational damages caused by a data breach or similar incident.

For this section, we can start by asking what the most common regulatory compliance laws are that organizations need to be aware of. But this question is very broad, and many regulations are industry-specific. In the following section, we will review some of the regulatory compliance requirements irrespective of the industry they apply to:

- **The Federal Financial Institutions Examinations Council (FFIEC):**

  - The FFIEC was established on March 10, 1979, under title X of the **Financial Institutions Regulatory and Interest Rate Control Act of 1978 (FIRA)**, Public Law 95-630. In 1989, title XI of the **Financial Institutions Reform, Recovery and Enforcement Act of 1989 (FIRREA)** established the **Appraisal Subcommittee (ASC)** within the Examination Council.

  - The Council is a formal interagency body authorized to direct consistent rules, standards, and report forms for the federal examination of financial institutions.

- **Healthcare Insurance Portability and Accountability Act (HIPAA):**

  - The HIPAA was signed into law on August 21, 1996.

  - On April 14, 2002, the HIPAA Privacy Rule defined **Protected Health Information (PHI)** as any information held by a covered entity that relates to the health status, the provision of healthcare, or the payment of healthcare that can be linked to an individual. The privacy rule also introduced 18 identifiers for PHI.

  - On April 21, 2005, HIPAA was enhanced to include the HIPAA Security Rule that dealt with **electronically stored Personal Health Information (ePHI)** and laid down the following three security safeguards: administrative, physical, and technical.

  - On March 16, 2006, HIPAA was enhanced by the introduction of the HIPAA Enforcement Rule that included the Department of **Health and Human Services (HHS)**, granting it the privilege to investigate covered entities that are not complying with HIPAA regulations. The Department's **Office for Civil Rights (OCR)** can prosecute criminal charges against persistent offenders, and individuals can also pursue civil legal action against covered entities.

  - In 2013, the HIPAA Final Omnibus (Final Rule) was enacted to include specific encryption standards to render ePHI unusable with new penalties and fines for non-compliance.

- **The Federal Information Security Management Act (FISMA)**:

  - FISMA is a US federal law, passed in 2002, that made it a requirement for federal agencies to develop, document, and implement an information security and protection program.

  - It applies to all agencies within the US federal government. The government expanded FISMA to include state agencies administering federal programs such as unemployment, insurance, Medicare, and Medicaid.

  - FISMA was created to require each federal agency to develop, document, and implement a complete information security plan to protect and support the operations of the agency.

- **HITECH and Breach Notification Rule**:

  - **Health Information technology for Economic and Clinical Health Act (HITECH)** was enacted in 2009.

  - In addition to covered entities, the HIPAA Rule applies to business associates as well.

  - The Breach Notification Rule:

    - All breaches of ePHI related to more than 500 individuals must be reported to the Department of Health and Human Services Office for Civil Rights.

  - **The Federal Risk and Authorization Management Program (FedRAMP)**:

    - FedRAMP began in 2011 as a way to ensure the security of cloud services used by the US government.

    - FedRAMP is mandatory for Federal Agency cloud deployments and service models at the low-, moderate-, and high-risk impact levels.

    - It offers a standardized approach to authorization, security assessment, and the continuous monitoring of cloud services and products.

    - Private cloud deployments intended for single organizations and implemented fully within federal facilities are the only exception.

  - **EU General Data Protection Regulation (GDPR)**:

    - The regulation was put into effect on May 25, 2018, and applies to EU citizens.

  - **The California Consumer Privacy Act of 2018**:

    - CCPA requires all businesses that collect personal information on California's residents to use *reasonable security procedures and practices appropriate to the nature of the information, to protect the personal information from unauthorized access, destruction, use, modification, or disclosure.*

- It includes a California resident's first name (or first initial) and last name coupled with sensitive personal information such as social security numbers, driver's license numbers, financial account numbers, and medical and health information.

Not all the preceding laws are applicable to all companies, but it is good to have an understanding of the various requirements for each industry.

> **Important note**
>
> ISACA does not ask specific questions related to an industry standard, framework, regulation, or geography.

In the next section, we will learn about the ethical challenges of risk management.

Ethics and risk management

Ethics are moral principles that drive an employee's judgment to perform daily activities and define socially acceptable behavior. Often, risk is impacted by professional ethics. It is easy to understand that an organization with poor ethical standards may be more susceptible to fraud or theft. Each organization has its own measures of maintaining ethical values and culture. For example, some organizations allow employees to receive gifts from clients and suppliers, but this is not acceptable at all for other organizations.

The risk of an employee violating the ethics policy of the enterprise can best be addressed by letting the senior management communicate the ethics policy to everyone and ensure that employees at all levels are appropriately trained on those policies.

Relationship between ethics and culture

Ethics and culture cannot be separated. Ethics is not a once-a-year *check-the-box* function, rather ethics must be inherent in an organization's culture. Without a strong culture, this check-the-box function could either be just training or a yearly seminar that employees are forced to attend, and once that is complete, the employees have no accountability for organizational culture and ethics.

> **Important note**
>
> For better or worse, culture starts at the top of the organization and is systemic. Good ethics must be an organization's cultural way of life.

Instead of organizations using phrases such as "Tone at the top," they should be focusing on "Tone from the top" to reflect the right ethics and behaviors for employees to follow and establish a culture of doing things right.

In the next section, we will learn about how ethics affect IT risk within an organization.

How do ethics affect IT risk?

Unlike in the past, when IT used to operate in complete silos and had a minimal effect on the rest of the organization, things have changed, and IT has become ingrained in each area of business. It is extremely prudent for the IT team to ensure that the right IT practices are followed across the organization without any exceptions.

Since IT teams have, historically, had more access than the rest of the employees so that they can perform maintenance activities and such, I have seen IT team members using those privileges to install malicious software for personal work. That malicious software installed viruses on their machine that then cascaded to other machines on the network. These types of incidents are not uncommon, but they cannot be completely eliminated unless the employees who are part of the IT teams follow the policies themselves.

It is important for organizations to make employees aware of the risk of breaching IT policies and how to report such incidents. The ethical issues in IT risk can best be managed by conducting mandatory annual ethics training and setting a Code of Ethics (also known as a Code of Conduct) for its employees. Additionally, organizations should provide an open line of communication such as anonymous emails, complaint hotlines, or surveys to employees so that employees can raise such concerns without the fear of ramifications.

In the next section, we will learn about the ISACA Code of Professional Ethics.

ISACA Code of Professional Ethics

ISACA has defined and set forth a code of professional conduct for members of the association, including CRISC holders and certified risk practitioners. ISACA certification holders shall do the following:

- Support the implementation of, and encourage compliance with, appropriate standards and procedures for the effective governance and management of enterprise information systems and technology, including audit, control, security, and risk management.

- Perform their duties with objectivity, due diligence, and professional care, in accordance with professional standards.

- Serve in the interest of stakeholders in a lawful manner, while maintaining high standards of conduct and character, and not discrediting their profession or the association.

- Maintain the privacy and confidentiality of information obtained in the course of their activities unless disclosure is required by a legal authority. Such information shall not be used for personal benefit or released to inappropriate parties.

- Maintain competency in their respective fields and agree to undertake only those activities they can reasonably expect to complete with the necessary skills, knowledge, and competence.

- Inform appropriate parties of the results of work performed, including the disclosure of all significant facts known to them that, if not disclosed, could distort the reporting of the results.

- Support the professional education of stakeholders in enhancing their understanding of the governance and management of enterprise information systems and technology, including audit, control, security, and risk management.

In the next section, we will review the summary of our learnings from this chapter.

Summary

At the beginning of this chapter, we learned about the major laws and regulations over a variety of industries and geographies that may pose a legal requirement for the organization to adhere to those regulations. Then, we learned about the relationship between ethics, culture, and IT risk management that is critical to determine an organization's response to risks. In the next section, we learned about the importance of professional ethics and ISACA's Code of Professional Ethics, which all the CRISC candidates and certification holders are expected to comply with.

In the next chapter, we will be diving into domain 2, *IT Risk Assessment*, and learn about the risk management life cycle.

Review questions

1. Which of the following is a federal law that provides guidance on protecting sensitive health information?

    A. CCPA

    B. HIPAA

    C. FFIEC

    D. GDPR

2. Which of the following groups is a beneficiary of CCPA law?

    A. US residents

    B. EU residents

    C. California residents

    D. Canada residents

3. According to the HIPAA Breach Notification Rule, which of the following is true and would require a **Covered entity** (**CE**) to report the breach to the OCR?

    A. Breach of more than 100 California residents' information

    B. IT incident that occurs in the CE

    C. IT incident that occurs in the business associate

    D. Breach of health information of more than 500 individuals

4. Failure to comply with the ISACA Code of Ethics could lead to which of the following?

    A. Mandatory training from ISACA

    B. An additional 20 questions in the exam

    C. Investigation and disciplinary measures from ISACA

    D. Immediate revocation of certification or membership

Answer

1. **B**. CCPA is related to the personal information of California residents, FFIEC is a banking regulation, and GDPR is related to the personal information of EU residents.

2. **C**. California residents is the correct answer. The other options are distractors.

3. **D**. The HIPAA Breach Notification Rule requires a notification to the OCR for the breach of health information of more than 500 individuals. The other options are distractors.

4. **C**. ISACA will conduct an investigation and take additional disciplinary actions upon failure to comply with the Code of Ethics. The other options are distractors.

Part 3:
IT Risk Assessment, Threat Management, and Risk Analysis

In this part, you will get a detailed understanding of the risk management life cycle; the relationship between threat, vulnerability, and risk; and how to develop a vulnerability management program. In addition, you will learn about risk assessment approaches, methodologies, frameworks, and techniques. Lastly, you will also get a detailed understanding of business impact analysis and inherent and residual risk.

This part has the following chapters:

- *Chapter 6, Risk Management Life Cycle*
- *Chapter 7, Threat, Vulnerability, and Risk*
- *Chapter 8, Risk Assessment Concepts, Standards, and Frameworks*
- *Chapter 9, Business Impact Analysis, and Inherent and Residual Risk*

6
Risk Management Life Cycle

This chapter marks the beginning of Domain 2, *IT Risk Assessment*, for CRISC. This domain represents 20 percent (approximately 30 questions) of the CRISC exam. As a reminder, Domain 1 of the CRISC exam and the material we learned up to *Chapter 5, Legal Requirements and the Ethics of Risk Management*, was entirely based on *Governance*, which relates to the direction from the stakeholders and leadership team. This chapter, and the following chapters, are about the hands-on approach to implementing those directions across the organization.

The aim of this chapter is to introduce the concept of risk, learn how it is different from IT risk, take a deeper dive into the risk management life cycle, understand the requirements of risk assessments, learn the difference between issues, events, incidents, and breaches, and ultimately, learn about how the correlation of events and incidents works. Additionally, we will learn about how to choose different sets of controls (detective/corrective/preventive) to influence the inherent risk and optimize the residual risk.

In this chapter, we will cover the following topics:

- Comparing risk and IT risk
- IT risk management life cycle
- Requirements of risk assessment
- Issues, events, incidents, and breaches
- Correlating events and incidents

With that, let us dive into the first section about risk and IT risk.

Comparing risk and IT risk

For any organization, **risk** could be the probability of having an adverse impact on the goals or outcome of an organization. As we learned in the earlier chapters, there could be risks related to geography, market, operations, finance, reputation, technology, natural disasters, and more.

IT risk is a subset of the overarching world of risk. It is the probability that a threat will exploit an information system vulnerability and could lead to the loss of IT systems, unauthorized disclosure/modification/destruction/loss of information, errors and omissions, or failure to run the operations successfully.

ISACA has also published a risk IT framework. It defines the IT risk for an organization as "*the business risk associated with the use, ownership, operation, involvement, influence, and adoption of IT within an enterprise. IT risk can be categorized into IT benefit/value enablement risk, IT programme and project delivery risk, and IT operations and service delivery risk.*"

An example of a geographical risk would be an office building prone to a natural disaster such as an earthquake, whereas the IT risk will be outdated systems without patches or any anti-malware solutions. For the purpose of this chapter and the CRISC outline, we will only be focusing on IT risks.

> **Important note**
>
> IT risk is a subset of enterprise risk, and the risk faced by an IT system is measured by the impact of an IT-related problem on the business services that the IT system supports.

In the next section, we will look at the IT risk management life cycle.

IT risk management life cycle

There are six steps in the IT risk management life cycle, starting from risk identification to risk monitoring. Each step is equally important to ensure that an organization does not only identify and assess the risk, but treats it according to its impact on the business and risk appetite, reports to the executives and key decision makers in an understandable report, and lastly, performs continuous monitoring of the risk and controls.

The following list details the key steps for IT risk management:

- **Risk identification**:

  Risk identification is the first step in IT risk management. An organization can only assess and treat the risk that it knows exists. Any failure to identify risks could lead to an organization not including those risks in its strategic planning and not giving them the due attention required.

  In IT risk management, it is important for the risk practitioner to be aware of the technologies used in the organization and understand the development, acquisition, implementation, integration, and sunset processes for those technologies.

- **Risk categorization**

  Once a risk has been identified, it is important to perform a risk analysis and categorize it accordingly. If the risk categorization is not done correctly, it will be extremely hard for IT risk practitioners to rate the risk as critical/high/medium/low and assign it to the correct stakeholders for further action.

- **Risk assessment**

  Risk assessment is the process used to evaluate risk based on likelihood and its potential impact on the critical functions necessary for an enterprise to continue business operations. Risk assessment helps an organization prioritize risks on the basis of their likelihood and potential impact. With the help of the risk assessment, risk practitioners can derive relationships between the risk and the enterprise's risk appetite and tolerance. Lastly, the risk assessment process provides information that is used to respond to risk in an appropriate and cost-effective manner.

- **Risk response and mitigation**

  Once the risk has been assessed and its severity is quantified, the organization can choose to respond to the risk by using any of the treatment strategies such as mitigating the risk by adding additional controls, accepting the risk if it falls within the risk appetite and the cost of control implementation is more than the cost of the asset, transferring the risk to reduce the liability such as by obtaining insurance for the asset, or completely avoiding the risk; for instance, shutting down the business process altogether as the cost of control implementation highly outweighs the cost of the assets or business benefits.

- **Risk reporting**

  Risk practitioners and business owners will have to obtain approval from the stakeholders on any risk response strategy they choose to employ for a risk. For that, it is important to present an appropriate business case to the senior management in the form of an executive report or dashboard that clearly shows the risks and control implementation strategy.

- **Risk and control monitoring**

  Risk management should always be an ongoing process for an organization. The threat landscape, threats, and vulnerabilities are changing at a faster rate than ever before, and any organization cannot take the risk of standing still with its risk management process. Therefore, it is important for an organization to monitor not just the risks but also the controls to ensure they are adequate with the changing landscape and that the risk level doesn't exceed the risk appetite.

The following figure shows the IT risk management life cycle:

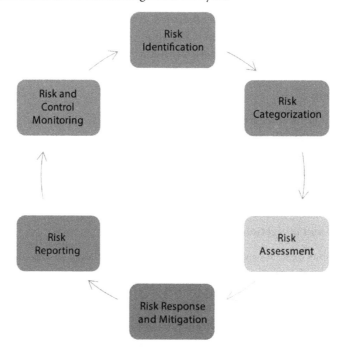

Figure 6.1 – IT risk management life cycle

Let us look at an example to concrete our understanding of the IT risk management life cycle.

When I worked for a large multinational bank, there were many incidents related to data leakage via email. Though employees were not allowed to access their personal email, they could attach files to an official email and send that out to their personal email. This posed a huge risk to the bank. Imagine an employee working on a banking application that has the name and account balance of all customers, and before leaving the company, they just decide to download the detailed list of all the account holders and send it to a competitor or, worse, hackers!

Historically, risk managers did not think of blocking the external email access of all employees, but after multiple similar incidents, the risk of data leakage via email was identified. This is known as **risk identification**, which is the first phase of the IT risk management life cycle.

The second phase of the IT risk management life cycle is **risk categorization**. So, after the initial identification, the team determined that the risk is a *Critical* category risk that needs to be prioritized and addressed immediately. There were several similar incidents, and compensating controls such as scanning emails sent outside the organization were not effective in mitigating the risk.

The third phase of the IT risk management life cycle is **risk assessment**, that is, determining the material impact of the risk. In this scenario, the team reviewed all the incidents of data leakage, the inherent risk to the organization due to lack of controls, and the residual risk after implementing additional controls such as a **Data Loss Prevention (DLP)** tool.

The next phase of **risk response and mitigation** deals with implementing the actual controls. Once the risk and controls are identified, it's time to act on them. It's critical to decide the type of mitigation that is required for this risk without being a bottleneck in the day-to-day role of employees who really need access to send emails, such as the marketing and communications team. At the time, we decided to block the external email access of all employees to reduce the number of incidents but also granted exceptions to employees who needed access for their roles. The exception was approved by the manager before being sent to the IT and security team with a deadline for when the exception could be revoked.

Following the risk mitigation phase, the next phase is **risk reporting** This involves visualizing all the incidents related to data loss due to emails across regions on a dashboard. Based on this, the DLP tool was implemented and the results were sent to the relevant stakeholders. The most important aspect of such a dashboard is to keep it actionable and as non-technical as possible, so stakeholders across the organization can understand it without needing to know the technical jargon.

The last phase of the IT risk management life cycle in this example is **continuous monitoring**. With the help of continuous monitoring of implemented controls, the team managed to answer questions such as are the implemented controls really effective? Were the number of external emails sent out with confidential information reduced? Are there any employees still trying to send confidential information to an external email address? Were the exceptions granted to particular employees revoked when they were no longer required?

In the next section, we will learn about the different laws and regulations that require that organizations conduct a risk assessment.

Requirements of risk assessment

Risk assessment is an important tool in the arsenal of risk practitioners. Risk assessments help organizations determine the level of risk and mitigate it according to the risk appetite of the organizations. Risk assessments also help organizations to be more proactive in implementing controls for unforeseen risks instead of being reactive to adverse risk scenarios.

The following table details some of the legal and regulatory compliance requirements for conducting a risk assessment:

Regulation/Law	Risk assessment requirements section
Canada – The **Personal Information Protection and Electronic Documents Act** (PIPEDA)	Principles 1,3, and 7
EU Directive 2016/679 – **General Data Protection Regulation** (GDPR)	Article 36. 1, 7(c), 7(d), and 11
Payment Card Industry Data Security Standard (PCI DSS)	v3.2.1, Requirement 12.2
US Federal Finance Institutions Examinations Council – IT Examination Handbook	Operations – Risk Management
US **Health Information Portability and Accountability Act** (HIPAA) Security Rule	45 CFR 164.308 (a)(1)(ii)(A)
US Nation Credit Union Administration – Guidelines for Safeguarding Member Information	7535-01-U, III.B Assess Risk
US – **New York State Department of Financial Services** (NY DFS) – Cybersecurity Requirements for Financial Services Companies	23 NYCRR 500.09
US Nuclear Regulatory Commission – Protection of digital computer and communication systems and networks	10 CFR 73.54
US Securities and Exchange Commission – Commission Statement and Guidance on Public Company Cybersecurity Disclosure	17 CFR Parts 229 and 249
US – The **Sarbanes Oxley Act** (SOX)	Section 404

Table 6.1 – Legal and regulatory compliance: risk assessment requirements

It is to be noted that not all the preceding laws prescribe the frequency of conducting risk assessments, but it is always a good practice to perform these assessments at least on an annual basis to comply with these requirements.

In the next section, we will look at some of the common terminologies that are required to be understood before moving on to the next section.

Issues, events, incidents, and breaches

This section is not specifically included in the CRISC exam syllabus, but it is important for an IT risk manager to understand certain terminologies. I have seen many experienced risk professionals use these terms interchangeably, but that's not correct. The following list details the definitions of each term:

- **Issues** – This is an instance of IT risk that has not materialized at all but needs to be considered and kept on the radar. This is a combination of control, value, and threat conditions that impose a noteworthy level of risk. One example of an issue could be outdated operating systems that are still being used by employees. Though nothing is wrong with using an outdated operating system and delaying the update by a few weeks while it's being tested, it should be noted as an issue, and the operating system should be updated at the earliest opportunity.

- **Events** – This is any occurrence that takes place during a certain period of time. As per ISACA, there can be three types of events: a threat event, a loss event, and a vulnerability event. An example of an event would be a failed login attempt. In this case, nothing is wrong with a single failed login attempt as long as it's logged and monitored. Users may enter an incorrect username or password that could lead to a login failure and, hence, an event.

- **Incidents** – An event or a combination of events that have a negative outcome affecting the confidentiality, integrity, or availability of an organization's data is considered an incident. An example of this would be thousands of failed login attempts within a second. Contrary to the preceding event example, a barrage of login requests in a short amount of time confirms a brute-force attack that should be responded to at the earliest opportunity.

> **Important note**
>
> An **event** is any occurrence that can be observed, verified, and documented, whereas an **incident** is one or more related events that negatively affect the company and/or impact its security posture.

- **Breach** – Any accidental or unlawful destruction, loss, alteration, unauthorized disclosure, or access of a user's data is considered a breach. The Yahoo breach in 2016 and the Equifax breach in 2017 are a couple of examples of data breaches.

In the next section, we will learn how to correlate events and incidents.

Correlating events and incidents

One of the major problems for any organization is the huge number of event alerts they receive. Over a period of time, it becomes difficult for the organization to maintain and correlate all the alerts.

IT event correlation automates the process of analyzing IT events and identifying relationships between them to detect problems and uncover their root cause. There are some event correlation tools such as AlertLogic, Splunk, and others that can help organizations monitor their systems and applications more effectively. This also helps to reduce false positives and improve uptime and performance.

IT infrastructures generate a huge amount of data in various formats. This could be from multiple sources such as servers, databases, virtual machines, mobile devices, operating systems, web applications, IoT devices, and other network components. An event for this kind of tool can be any piece of data that provides insight about a state change in that infrastructure, such as a successful or failed user login. Many of these events are normal and do not require an immediate response, but some will signify a problem within the infrastructure that, as we learned earlier, are called incidents. An organization can generate and process thousands of events each day, and making sense of them to determine which are relevant represents a significant challenge for IT teams.

The IT event correlation software mentioned earlier ingests data from the available sources and uses machine learning to recognize meaningful patterns, relationships, and inferences. These techniques enable teams to more easily identify and resolve incidents and outages, conduct performance monitoring, and help improve the availability and stability of the infrastructure.

Summary

At the beginning of this chapter, we learned about risk and how it differentiates from IT risk. Then, we learned about the IT risk management life cycle and understood the process in detail with the help of an example. We then learned about the legal and compliance requirements of conducting a risk assessment. In the next section, we switched gears to learn about the difference between issues, events, incidents, and breaches and looked at an overview of event correlation.

In the next chapter, we will learn about the fundamentals of risk, that is, threats and vulnerabilities, and how they relate to risk.

Review questions

1. Which of the following depicts the correct relationship between IT risk and enterprise risk?

    A. Enterprise risk is a part of IT risk.

    B. IT risk is a part of enterprise risk.

    C. These types of risk are not related to each other.

    D. Enterprise risk is independent of IT risk.

2. Which risk management life cycle step emphasizes the "You can't protect what you don't know exists" dictum?

   A. Risk and control monitoring

   B. Risk assessment

   C. Risk identification

   D. Risk categorization

3. Which of the following is a formal requirement by many legal and regulatory compliance frameworks?

   A. Performing vulnerability assessments

   B. Maintaining an asset inventory

   C. Deploying changes to production without testing

   D. Conducting a formal risk assessment

4. A healthcare employee accidentally sent more than 5,000 patient records in response to a phishing email. The patient records are not encrypted at rest or in transit. This scenario will fall under a material _____.

   A. Issue

   B. Event

   C. Incident

   D. Breach

5. What is the primary purpose of a SIEM?

   A. Reduce false positives as much as possible

   B. Ingest all the data it possibly can and create beautiful dashboards

   C. Correlate events and alert on actionable insights

   D. Provide API integration with other systems

Answers

1.   **B**. IT risk is part of enterprise risk. The other options are distractors.

2.   **C**. This is the requirement of the risk identification step.

3.   **D**. Conducting a formal risk assessment is a requirement of many legal and regulatory compliance frameworks. The other options are important for a risk management program but are not mandatory.

4.   **D**. This scenario will fall under a breach since the material data, for instance, the PHI of patients is leaked.

5.   **C**. All the options are applicable, but the primary objective of a SIEM is to correlate events from the system and alert on malicious activities by providing actionable insights.

7

Threat, Vulnerability, and Risk

Threat, **vulnerability**, and **risk** are three important concepts that are required to understand the risk management life cycle concretely. Risk practitioners must know these concepts off the top of their heads as they come in extremely handy at the time of risk assessment and threat modeling, both of which we'll learn about later in this book.

This chapter aims to introduce the concepts of threat, vulnerability, and risk, understand the relationship between each, and learn about threat modeling and the threat landscape. We will also learn about vulnerability and control analysis and vulnerability sources, and briefly touch on building a vulnerability management program.

In this chapter, we will cover the following topics:

- Threat, vulnerability, and risk
- The relationship between threat, vulnerability, and risk
- Understanding threat modeling
- Vulnerability analysis
- Tools for identifying vulnerabilities
- Vulnerability management program

With that, let's dive into the first section: *Threat, vulnerability, and risk*.

Threat, vulnerability, and risk

Like all the previous chapters, we will start this chapter by learning about the definitions of threat, vulnerability, and risk.

A **threat** could be anything (a human, malicious code, a bot, a natural disaster, and so on) that could impact an asset and adversely affect it in a manner that can result in harm. Threats employ threat actors to exploit a vulnerability and a threat vector is the path or route that's used by the adversary to gain access to the target.

A **vulnerability** is a weakness in the design, implementation, operation, or internal control of a process, which could expose the system or an asset to adverse threats from threat events.

When a threat exploits a vulnerability and adversely affects the system, it is considered a **risk**.

It is important to note that threats will always exist and there is little that an organization can do to limit the number of threats. However, organizations can always choose to apply sufficient controls that could limit the probability of a threat exploiting a vulnerability and resulting in risk.

> **Important note**
> It's essential to understand that threats are not within our control. However, we can control vulnerabilities and establish measures to address those vulnerabilities.

As we learned in the previous chapter, the goal of a risk manager is to minimize risk, and one of the first steps to do this is to identify all the relevant risks. The first step to do so would be to document all the relevant threats to the organization that could either relate to technology or non-technology and then identify the corresponding vulnerabilities – that is, any deficiencies in the organization's system that could be exploited. The next step would be to ensure that sufficient controls are in place to mitigate the impact of the vulnerabilities being exploited to an acceptable level.

In the next section, we will take a more in-depth look into the relationship between these terms.

The relationship between threats, vulnerabilities, and risk

A risk manager needs to understand the relationship between threat, vulnerability, and risk. In addition, a risk manager should also understand the impact of threat actors and threat vectors and how they result in risk to assets.

Any threat by itself could not result in risk. It needs a vulnerability that it can exploit to cause risk to the system. A threat also needs a threat actor, which will materialize the threat by using a threat vector. The threat vector will then materialize the vulnerability and cause risk, which will harm the asset.

The following figure shows the relationship between the key concepts of threats, vulnerabilities, risks, and assets:

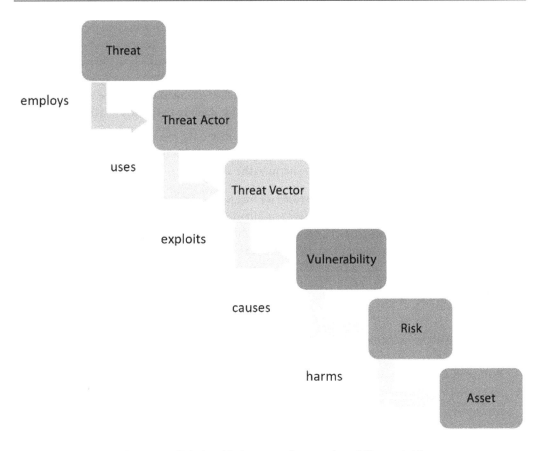

Figure 7.1 – Relationship between threat, vulnerability, and risk

Let's detail this with the help of an example. The scenario in our example is of malicious software being installed and propagated on all the machines on a network:

- **Threat**: In this scenario, the threat will be the malicious actor that wrote the malicious software in the first place. If you are not familiar, all the viruses that we see on our machines or popups with ads that lead to malicious sites are written by some malicious actor that could either be doing it for fun or with malicious intent. That malicious actor is the threat.

  The identified threat may employ a threat actor to spread the virus to other systems. This threat actor could be a person sending malicious files to random individuals over emails, a malicious website that would lure victims to download malicious files, or they could attach malicious software with a legitimate file such as a PDF or Word file. These PDF/Word files, malicious websites, or random emails are the threat vectors that would exploit the vulnerabilities in a system to install malicious software.

- **Vulnerability**: In this scenario, the vulnerability could be that the person hosting the website did not do a routine hygiene check or did not have the controls to stop the malicious actor before uploading the file. On the host side, the vulnerability could be that the host did not install an antivirus solution that could prevent the file from opening in the first place or an email scanner that could scan the files sent by some random individual on the internet. All these vulnerabilities have controls and can be fixed before they cause undue damage to the host machines.

- **Risk**: Once these vulnerabilities have been exploited – that is, the malicious file has been opened on the host machine – the virus will spread on all the machines, putting the machines at risk of malware infection and hence causing harm to the asset.

This was a generic scenario to detail the concepts we touched on at the beginning of this chapter. In the real world, there are thousands of threats and thousands of vulnerabilities that are fixed and generated at the same time. Organizations are always on the lookout to find the next *perfect* solution that could address all threats and vulnerabilities, but alas, the threat landscape is always changing and there is not – and will not be – a single solution that could act as a magic potion to fix systems.

Any organization that is daunted by these threats and threat mitigation could start with the following key steps:

1. **Be proactive**: Organizations can always be more proactive and implement controls to reduce significant risk to critical systems. There are many sources of vulnerabilities, such as the **National Vulnerability Database** (**NVD**), that organizations can routinely check or get alerts from for critical vulnerabilities reported across the industry and implement controls before being exploited. Organizations with sufficient budgets should invest in hiring a dedicated security researcher to identify the vulnerabilities before attackers do and/or develop a bug bounty program to let security researchers from across the world report vulnerabilities responsibly.

2. **Prioritize**: Not all vulnerabilities impact an organization and not all of them will be of critical or high priority that need to be remediated right away. The best way to prioritize these vulnerabilities would be to perform a threat and vulnerability analysis and ensure that adequate controls are implemented for each vulnerability.

We will look into this in detail toward the end of this chapter. In the next section, we will learn about threat modeling and the threat landscape.

Understanding threat modeling

Threat modeling is a structured approach to identifying threats, potential vulnerabilities, and corresponding security requirements, quantifying threat and vulnerability criticality, and prioritizing remediation per severity. It is performed as a proactive measure during product design and development to ensure that adequate controls are in place before the deployment.

There are four generic steps in threat modeling:

1. **Model**: What are we building?
2. **Identify Threats**: What could go wrong?
3. **Mitigate**: What countermeasures do we have to defend against the threats?
4. **Validate**: Have we performed all the previous steps?

These steps can be visualized as follows:

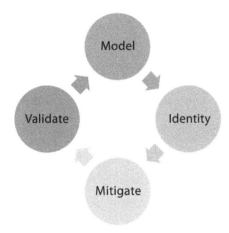

Figure 7.2 – The threat modeling cycle

The purpose of threat modeling is to provide defenders with a systematic analysis of the most likely attack vectors. Threat modeling helps in identifying high-value assets, assets that are vulnerable to attack, and possible methods of attack.

Threat modeling methods

There are seven well-known methods of threat modeling. We will go through each of them in this section:

- **Security development life cycle (SDL)**: The goal of this model is to introduce security at every stage of software development to reduce the number of security-related design and coding defects, as well as to reduce the severity of any defects that are left. Aptly, the motto of the SDL is **Secure by Design, Secure by Default, Secure in Deployment and Communication (SD3+C)**.

- **STRIDE**: STRIDE stands for **Spoofing**, **Tampering**, **Repudiation**, **Information Disclosure**, **Denial of Service**, and **Elevation of Privilege**. STRIDE is a model of threats and helps find threats to a system. Each property of STRIDE aims to support a desired security principle:

Threat	Desired Security Principle
Spoofing	Authenticity
Tampering	Integrity
Repudiation	Non-repudiation
Information Disclosure	Confidentiality
Denial of Service	Availability
Elevation of Privilege	Authorization

Table 7.1 – Security principles supported by the STRIDE framework

- **DREAD**: This model helps us assess risk using the five factors that affect it:

  - **Damage**: How bad is an attack?

  - **Reproducibility**: How easy is it to reproduce the attack?

  - **Exploitability**: What level of effort is required to launch the attack?

  - **Affected users**: How many people will be impacted?

  - **Discoverability**: How easy is it to identify the threat?

- **Process for Attack Simulation and Threat Analysis** (**PASTA**): PASTA is a threat modeling methodology centered around risk analysis, enabling organizations to integrate security strategy effectively right from the start. By emphasizing collaboration among all stakeholders, PASTA fosters a security-focused environment that incorporates contextual insights. There are seven stages for PASTA threat modeling:

  A. **Define the objectives**: This helps set the tone for the project or assessment.

  B. **Define the technical scope**: This defines the boundaries of the project.

  C. **Application decomposition**: This allows you to map what is important to what is in scope.

  D. **Threat analysis**: Here, you can document relevant threats and relate threat patterns to the data gathered.

  E. **Vulnerability and weakness analysis**: Here, you can examine the threats, rank them, and assess their likelihood.

  F. **Attack modeling**: Here, you assess the threats and turn them into attacks by examining the attack surface before and after the proposed changes.

  G. **Risk and impact analysis**: This involves developing the rationale for mitigation based on residual risk.

1. **Trike**: Trike is a security audit framework that employs threat modeling as a core technique and takes a risk-management and defensive approach.

2. **Visual, Agile, and Simple Threat (VAST) methodology**: VAST represents a significant departure from the limitations and hurdles that were encountered in existing threat modeling methodologies. Its core principle is rooted in the notion that for threat modeling to deliver tangible results, it must scale across the infrastructure, integrate seamlessly into Agile environments, and deliver actionable, precise, and consistent outputs that cater to developers, security teams, and senior executives.

3. **Operationally Critical Threat, Asset, and Vulnerability Evaluation (OCTAVE)**: OCTAVE primarily concentrates on evaluating organizational risks, omitting technological risks from its scope. Its key facets include operational risk, security practices, and technology.

In the following section, we will learn about the importance of threat modeling and the reasons to prioritize it.

The importance of threat modeling

As we learned earlier, there are many threat modeling methods that organizations can choose based on their requirements and specific use case. However, the result of each threat modeling exercise should result in the identification of specific threats and the ability to lessen the impact of threats that cannot be avoided.

A threat modeling exercise allows the organization to identify and address the biggest threats and plan mitigations for identified and documented threats. Threat modeling also allows developers to remediate security issues early in the SDL, which is extremely important to reduce the cost of remediation and speed up the deployment. Lastly, threat modeling helps in elevating the security posture of the application cost-effectively, prioritizing development and testing efforts based on threats identified in the exercise, and ultimately calculating and reducing the impact of residual risk.

Vulnerability analysis

A risk manager must dedicate sufficient time and resources to ensure the vulnerabilities identified using threat modeling are actioned. The first step that must be performed after identifying these vulnerabilities is to ensure that vulnerabilities are categorized per their severity. Analysis must also be performed on the implemented controls.

Organizations may choose to adopt a nomenclature to define the severity of the vulnerability. As an industry practice, vulnerabilities are quantified as *Critical*, *High*, *Medium*, *Low*, and *Informational*. Organizations and vulnerability assessment tools use the **Common Vulnerability Scoring System (CVSS)** to quantify vulnerabilities. A risk manager must prioritize the *Critical* and *High* vulnerabilities to be remediated as soon as practical and analyze the implemented controls periodically.

Tools for identifying vulnerabilities

Many vulnerability assessment tools on the market focus specifically on finding vulnerabilities; however, a risk manager must be aware of all the tools and resources that could be leveraged as a source for vulnerabilities. The following are some resources that could be used by a risk manager to surface vulnerabilities:

- **Vulnerability assessment scans**: Vulnerability assessment tools such as Nessus and Qualys could be a good source of information for open vulnerabilities.

- **Penetration tests**: It is standard for organizations to perform a penetration test at least annually or after a major change in the underlying infrastructure and systems. The findings from the penetration test are a good source for identifying vulnerabilities.

- **Static analysis**: Issues flagged by the static analysis tools in a code pipeline are a good source for finding vulnerabilities. The majority of the issues that are flagged by these tools are related to logical issues related to the code.

- **Dynamic analysis**: Dynamic scanning tools scan the code while it's running. They focus more on vulnerabilities that could arise while the application is running. Unlike static analysis tools, dynamic analysis tools do not have access to the program's source code.

- **Configuration checks**: There might be hundreds of engineers working on a cloud environment for a company. Each of these engineers might be spinning up an infrastructure for the modules they are working on. This exposes the organization to vulnerabilities that could occur due to misconfiguration issues. Performing a periodic configuration scan can provide a fair assessment of configuration issues.

- **Risk assessments**: Historically, risk assessments are not considered a source of vulnerability given that they primarily focus on finding non-technical risks. However, with the new methodologies of building code, such risks must be considered vulnerabilities.

- **Zero-day findings**: Security researchers publish zero-day vulnerabilities – that is, vulnerabilities that haven't been discovered yet and a patch is not available. A risk manager should follow these 0-day reports and ensure that the organization takes measures to fix them as early as possible if they are impacted.

- **Industry advisories**: Government organizations such as the **National Institute of Standards and Technology (NIST)** and the **Cybersecurity and Infrastructure Security Agency (CISA)** publish advisories on recent threats and vulnerabilities. Risk managers should subscribe to such advisories and keep themselves abreast of the latest findings in the industry.

- **Vendor security feeds/bulletins**: Vendor security feeds and bulletins are information resources provided by technology vendors to keep their customers informed about security threats and vulnerabilities that may impact their products or services.

There could be many other sources of vulnerabilities per the design and implemented tools. The preceding categories are a snapshot of the tools that are commonly used in practice as a source of vulnerability.

Vulnerability management program

Identifying vulnerabilities in itself will not make the organization secure but prioritizing and remediating those vulnerabilities per their severity will. A risk manager's primary job related to vulnerabilities is to ensure that the vulnerabilities are prioritized, tracked, and fixed as a part of the **vulnerability management program** (**VMP**).

Organizations may choose to implement a tool to input all the vulnerabilities from the sources mentioned in this chapter or manage them manually in a project management tool. An important aspect to note for a VMP is coordinating with other teams. Remediating a vulnerability may come across as additional work on top of an engineer's day-to-day; however, it is important to carve out some time and ensure that these vulnerabilities are remediated in an agreed-upon timeline.

Summary

At the beginning of this chapter, we learned about threat, vulnerability, and risk and their relationship with each other. We then learned about threat modeling, the importance of threat modeling in the software development life cycle, and the important threat modeling practices used in the industry. After, we switched gears to learn more about vulnerabilities, vulnerability and control analysis, sources of vulnerabilities, and how to start a vulnerability management program from scratch.

In the next chapter, we will learn about risk assessment concepts, standards, and frameworks.

Review questions

1. Which of the following is considered a weakness that could be exploited by a malicious actor?

    A. Threat

    B. Threat actor

    C. Vulnerability

    D. Risk

2. Which of the following is the *best* phase to perform threat modeling?

    A. Requirements

    B. Design

    C. Development

    D. Testing

3. *What could go wrong?* corresponds to which phase of threat modeling?

    A. Model

    B. Identify

    C. Mitigate

    D. Validate

4. Which of the following vulnerabilities should be prioritized for remediation?

    A. Low

    B. Medium

    C. High

    D. Critical

5. Which scoring technique is used to quantify a vulnerability?

    A. CSS

    B. CSV

    C. CVSS

    D. CVVS

Answers

1. **C.** Vulnerabilities are considered weaknesses that can be exploited by malicious actors.

2. **B.** It is important to perform threat modeling in the earlier stages of SDLC to ensure that threats are mitigated promptly. Threat modeling cannot be performed in the *requirements* phase as the team is still gathering the requirements and is not sure what the end product will look like.

3. **B.** The *identify* phase of threat modeling requires us to ask "What could go wrong?" and identify the corresponding threats.

4. **D.** Critical vulnerabilities pose extreme risks to the organization and should be prioritized for remediation.

5. **C.** The **Common Vulnerability Scoring System** (**CVSS**) is used to quantify vulnerabilities.

Risk Assessment Concepts, Standards, and Frameworks

In this chapter, we will aim to continue the learnings from the previous chapter where we learned about threats, vulnerabilities, and how they translate into risk. In this chapter, we will dive deep into **risk assessment concepts** (risk scenarios, risk register, and so on), **risk assessment standards**, and **risk management frameworks**. We will also learn about maintaining an effective risk register and how we can leverage already available industry risk catalogs to baseline the risk assessment program for an organization.

In this chapter, we will cover the following topics:

- Risk assessment approaches
- Risk assessment methodologies
- Risk assessment frameworks
- Risk assessment techniques
- Importance of a risk register

With that, let us dive into the first section on risk assessment approaches.

Risk assessment approaches

There are two approaches to risk management—top-down and bottom-up.

In a **top-down risk assessment approach**, risk scenarios are driven from the management perspective that are related directly to the organization's business objectives. In this approach, risk scenarios are developed for risk events that directly impact the business goals and objectives. Actions from the top-down risk assessment are easier to have a buy-in from the other stakeholders given these are directed from the senior management. The results of the risk assessment tend to be broader in nature as these are based on the experience of the management team managing multiple businesses and functions.

Here's an example of a top-down risk assessment for a healthcare organization's new **Electronic Health Record** (**EHR**) system, with the **Board of Directors' (BoD)** review of the quarterly revenue results, **Ethics & Compliance** (**E&C**), senior management performance, and security reports as crucial considerations:

- The BoD considers strategic risks with revenue implications and how the EHR aligns with organizational goals

- E&C reviews regulatory risks, ensuring adherence to health data laws

- Senior management evaluates departmental risks, considering specific functions such as outpatient care, pharmacy, and billing

- Cyber reports inform the assessment of system-specific risks such as data security and cyber-attack susceptibility

Each level involves identifying risks, estimating impact, and developing mitigation strategies, ensuring alignment with the organization's overarching goals.

The top-down approach ensures that the organization aligns its risk management strategies with broader organizational goals, and provides a holistic view of risk that can be used to guide decision-making.

In a **bottom-up risk assessment approach**, risk scenarios are identified by individuals and teams that are then cascaded up to the department, **business unit** (**BU**), and organization level. The risk scenarios identified in a bottom-up approach are more catered toward a particular business or function risk as these are raised by individuals who are experts in a particular field.

Let's consider the example we just discussed in the context of a bottom-up risk assessment for the new EHR system:

- The process begins at a granular level. System-level components such as data storage, access controls, logging and monitoring, cyber vulnerabilities, and so on are analyzed first.

- Next, individual departments such as patient care, pharmacy, and billing assess the EHR's operational impact and associated risks.

- Senior management then understands how these identified risks influence the performance and achievement of **key performance indicators** (**KPIs**).

- Finally, at the organizational level, E&C and the BoD consider these identified risks, evaluating the potential impact on regulatory compliance, quarterly revenue results, reputational risk, and alignment with strategic goals.

The bottom-up approach ensures a detailed understanding of the system and operational risks, connecting them to larger organizational objectives for a comprehensive risk profile.

The following diagram shows top-down and bottom-up risk assessment approaches:

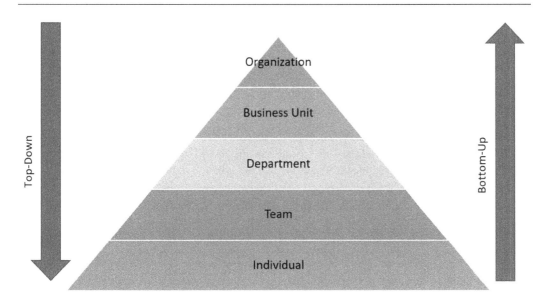

Figure 8.1 – Risk assessment approaches

Top-down risk governance is led by executives and leverages their strategic insight to identify and manage risks. This approach is easier due to fewer decision-makers; however, it tends to be exclusive and can potentially overlook crucial operational risks.

Contrarily, bottom-up risk governance involves *all* organizational levels, providing a comprehensive risk profile due to the diverse perspectives. Employees on the front lines contribute valuable insights on specific process-related risks. However, managing this inclusive approach is challenging due to the large number of participants, complex communication, and the need to align differing risk perspectives with overarching strategic and organizational goals.

Which is the best approach?

For best results, an organization should include both the top-down and bottom-up approaches for developing risk scenarios. Using both approaches simultaneously ensures that major risks to business objectives are addressed by the management team, whereas people, process, and technology risks are addressed by individuals and teams.

In the next section, we will cover the primary risk assessment methodologies.

Risk assessment methodologies

An IT risk manager should be hands-on when it comes to performing a **risk analysis**. The results of a risk analysis directly impact the risk response and, consequently, the resources allocated to each risk area. Therefore, the risk manager should be able to guide and perform the risk analysis and propose a risk response.

There are two primary types of risk analysis—**qualitative risk analysis** and **quantitative risk analysis**.

As the name suggests, qualitative risk analysis is based on qualitative parameters such as *High*, *Medium*, *Low*, and *Very Low* to depict the level of risk. These parameters are assigned to each risk scenario according to their likelihood and impact based on the experience and expertise of the group conducting the risk analysis and therefore may result in subjective outcomes.

On the contrary, quantitative risk analysis is more measured and aims to provide the monetary value at risk if the risk scenario materializes. This type of risk analysis is dependent on historical data for modeling, simulating, and calculating the likelihood of a risk scenario materializing. The risk manager needs to ensure that the data used for inferences for the quantitative risk analysis is *reasonably accurate* so that the resulting model that heavily relies on this data is equally accurate as well.

The following table shows the differences between a qualitative and quantitative risk assessment:

Qualitative	Quantitative
Subjective	Objective
Risk severity is important	Monetary value is important
Requires experience and expertise	Requires historical data and modeling
Relatively straightforward	Complex; may require additional tools and computation
Results can be biased based on the experience and expertise of the team	Results are completely based on data
Less expensive	More expensive; additional calculation and modeling tools may be required

Table 8.1 – Differences between qualitative and quantitative risk analysis

It is difficult for organizations to rely solely on qualitative or quantitative risk analysis, and therefore they can use a third approach to risk analysis that includes components from each. This approach to risk analysis is called **semiquantitative** or **hybrid risk assessment**.

Hybrid risk assessment combines the value of qualitative and quantitative risk assessment. It is a hybrid approach that considers the input of a qualitative approach combined with a numerical scale to determine the impact of a quantitative risk assessment. In a hybrid risk assessment, each risk rating is assigned a score—say, *1-5*, with *1* being the lowest likelihood/impact and *5* being the highest likelihood/impact. The results of the hybrid risk assessment could range from *1* to *25* (*likelihood x impact*) and provide the right guidance on what should be the focus for risk response.

A risk practitioner should always aim to strike a balance and choose the method that is most appropriate for the organization. If an organization is in the infant stages of the risk assessment program, a qualitative risk assessment exercise will be helpful to hit the ground running and initiate a risk assessment program, whereas, for an organization that is comfortable conducting a qualitative risk assessment exercise, the risk manager should aim to graduate them to either a hybrid or a quantitative risk assessment.

In the next section, we will review the risk assessment frameworks that are already available to perform risk assessment.

Risk assessment frameworks

There are many industry-standard risk assessment frameworks that a risk manager can choose to perform a risk assessment. The risk manager is responsible for ensuring that the organization utilizes the framework that makes the most sense for its risk assessment.

The following is a summary of common industry risk assessment frameworks:

Framework	Description
NIST SP 800-30	Risk management for general information systems
NIST SP 800-37	Risk management for federal information systems
NIST SP 800-161	Risk management for supply chain management
ISO/IEC 27005	Risk management for information systems
ISO/IEC 31010	Risk management for IT governance
ISO/IEC 31000	Organizational risk management
British Standard 100-3	Risk analysis based on IT infrastructure
OCTAVE	Operationally critical, threat, asset, and vulnerability evaluation for enterprise IT projects
FAIR	Quantitative risk management
ISACA IT Risk	Risk management for enterprises to manage IT risk

Table 8.2 – Risk assessment frameworks

The preceding frameworks are the industry standards for risk management. The risk manager should opt to use the framework that makes the most sense per the requirements of the organization.

In the next section, we will learn how to perform risk analysis and start building a **risk register**.

Risk assessment techniques

Conducting a risk assessment is one of the primary skills that each risk practitioner should learn. The results of a risk assessment help the management to prioritize the risk and decide on appropriate risk response strategies.

There are numerous techniques for performing a risk assessment, and none of the techniques is completely right or wrong. The risk practitioner should choose the training they feel is appropriate for their organization and makes the most sense to the stakeholders.

Here is a summary of major risk assessment techniques:

- **Bayesian analysis**: A statistical inference that uses prior distribution data to determine the probability of a result. This technique relies on the prior distribution data to be accurate in order to be effective and produce accurate results.

- **Bow-tie analysis (BTA)**: Provides a diagram to communicate risk assessment results by displaying links between possible causes, controls, and consequences.

- **Brainstorming/interview**: This technique gathers a large group of types of potential risks or ideas to be ranked by a team. The initial interview or brainstorming is often completed using prompts or interviews with an individual or a small group.

- **Cause and consequence analysis**: This analysis combines techniques of **fault tree analysis (FTA)** and **event tree analysis (ETA)** that allow for time delays to be considered.

- **Cause and effect analysis**: This analysis looks at the factors that contributed to a certain effect and groups the causes into categories, which are then displayed using a diagram.

- **Checklists**: A checklist is a list of potential or typical threats or other considerations that should be of interest to the organization, and these items can be checked off as completed.

- **Delphi method**: This method leverages expert opinion received using two or more rounds of questionnaires. After each round of questioning, the results are summarized and communicated to the experts by a facilitator. This collaborative technique is often used to build a consensus among experts.

- **ETA**: This is a forward-looking, bottom-up model that uses inductive reasoning to assess the probability of different events resulting in possible outcomes.

- **Factor Analysis of Information Risk (FAIR)**: FAIR is a taxonomy of the factors that contribute to risk and how they affect each other. It is primarily concerned with establishing accurate probabilities for the frequency and magnitude of data loss events.

- **FTA**: This technique starts with an event, examines possible means for the event to occur (from top to bottom), and displays these results in a logical tree diagram. This diagram can be used to generate ways to reduce or eliminate potential causes of the event.

- **Human reliability analysis (HRA)**: HRA examines the effect of human error on systems and their performance.

- **Lotus blossom method of brainstorming**: This is a brainstorming technique that makes use of visual representations by adding the problem in the center of the *blossom* of a 3×3 square. Then, related solutions or themes are added in an iterative manner, expanding outward.

- **Markov analysis**: This is used to analyze systems that can exist in multiple states. The Markov model assumes that future events are independent of past events.

- **Monte Carlo analysis**: This is a risk management technique that is used for conducting a quantitative analysis of risks. This technique is used to analyze the impact of risks on your project. Monte Carlo methods, or Monte Carlo experiments, are a broad class of computational algorithms that rely on repeated random sampling to obtain numerical results.

- **Operationally Critical Threat, Asset, and Vulnerability Evaluation (OCTAVE)**: OCTAVE is a flexible and self-directed risk assessment methodology where business and IT teams work together to address the security needs of the enterprise.

- **Sneak circuit analysis (SCA)**: This technique is used to identify design errors or sneak conditions that are often undetected by system tests.

- **Structured what-if technique (SWIFT)**: This technique uses structured brainstorming to identify risk. It uses prompts and guide words that are used with other risk analysis and evaluation techniques.

For an organization performing risk assessment for the first time, a quantitative technique such as FAIR will not be the most helpful. Similarly, for an organization that has performed multiple risk assessments and has graduated to a *mature* state of risk assessment, a **checklist-based risk assessment** will not be sufficient. Therefore, an organization's choice of risk assessment methodology should align with its maturity and experience in conducting risk assessments.

> Important note
>
> An organization should choose the risk assessment methodology that makes the most sense and aligns with the risk culture. The risk assessment can be performed as frequently as quarterly and must be performed at least annually.

In the next section, we will learn about the importance of maintaining an effective risk register after the risk assessment is complete.

Importance of a risk register

All risks identified in the risk assessment should be entered into a risk register, which could be a sophisticated **Software-as-a-Service (SaaS)** program or a spreadsheet. At a minimum, the risk register is to maintain details of threats, vulnerabilities, likelihood, impact, inherent risk, current controls, residual risk, countermeasures that will reduce the risk in the future, and a risk owner.

Not all the risks captured in the risk register will have the same priority, and a risk practitioner should dedicate sufficient time to determine which risks should be prioritized for remediation and which should be revisited later.

The best way to identify the risks that should be prioritized is to discuss the likelihood and impact with the stakeholders at the time of risk assessment. This helps the risk manager to eliminate guesswork on which risks could cause more damage to the organization and should be remediated on priority. If a risk cannot be remediated immediately, a **corrective action plan (CAP)** should be created for future remediation. The CAP should have an owner, a timeline in which the risk will be remediated, and the corrective actions that will be taken as part of the remediation. Once the remediation is implemented, the CAP can be closed and the risk can be marked as remediated.

Summary

At the beginning of this chapter, we learned about risk assessment approaches and the best ways to address risks. We then learned about risk assessment methodologies and industry-wide risk assessment frameworks. Each organization has unique requirements for risk assessment and each risk analysis needs a different skill set; therefore, the risk manager needs to understand the current maturity and choose a risk analysis technique that will be suitable for them. In the next section, we switched gears to learn more about the actual risk assessment techniques that the organization can use and the importance of maintaining an effective risk register after the risk assessment is complete. The risk manager can opt for different techniques before settling on one or perhaps use a combination of these techniques to understand the risks from different perspectives. The risk register should be a live document and should be updated whenever the risk changes due to internal or external factors.

In the next chapter, we will learn about **business impact analysis (BIA)**, how is it different from risk assessment, important concepts related to BIA, and inherent/residual/current risk.

Review questions

1. Which of the following frameworks is primarily used for quantitative risk management?

    A. NIST *800-30*

    B. FAIR

    C. ISO *27001*

    D. ISO *27005*

2. A top-down risk assessment starts from the __.

    A. Team

    B. Individual

    C. Organization

    D. Department

3. A bottom-up risk assessment starts from the __.

    A. Team

    B. Individual

    C. Organization

    D. Department

4. Which of the following is *NOT* true about qualitative risk management?

    A. Less expensive

    B. Subjective

    C. Requires complex computation

    D. Focused on severity

5. Which of the following NIST frameworks provides guidance for supply chain management?

    A. *800-161*

    B. *800-30*

    C. *800-57*

    D. *27001*

6. Which of the following risk assessment techniques provides the results of a risk assessment by displaying links between possible causes, controls, and consequences in terms of a diagram?

   A. FAIR

   B. BTA

   C. Markov analysis

   D. Monte Carlo analysis

7. The results of a risk assessment should be summarized as a(n) __.

   A. CAP

   B. Business continuity plan

   C. Organizational chart

   D. Risk register

Answers

1. **B.** FAIR is primarily used for quantitative risk management; all the other options are for qualitative risk management.

2. **C.** A top-down risk assessment starts from the organization and moves to the individual level.

3. **B.** A bottom-up risk assessment starts from the individual and moves to the organizational level.

4. **C.** Qualitative risk management doesn't require complex computations. Quantitative risk management requires complex computations.

5. **A.** NIST *800-161* is used for supply chain management. The other options are distractors.

6. **B.** BTA shows the results of a risk assessment in the form of a diagram.

7. **D.** A risk register is the immediate result of a risk assessment. All the other options are distractors.

9

Business Impact Analysis, and Inherent and Residual Risk

The aim of this chapter is to detail the differences between **business impact analysis (BIA)** and **risk assessment**, learn concepts that are related to BIA, understand the differences between inherent and residual risk, and finally, review how BIA can be used to ensure **business continuity (BC)** and effective disaster recovery planning.

In this chapter, we will cover the following topics:

- Differentiating between BIA and risk assessment
- Key concepts related to BIA
- Understanding types of risk

With that, let us dive into the first section, in which we will understand the differences between BIA and risk assessment.

Differentiating between BIA and risk assessment

BIA and risk assessment are related terms but not the same. In my experience, many practitioners use the terms interchangeably, which is incorrect.

BIA is the process of identifying critical business processes for an organization by assessing the impact of a disaster on that process. The primary objective of BIA is to determine systems, processes, or tools that will impact the identified business process in a positive or negative manner and then prioritize the recovery of business-defined critical services that support strategic objectives and goals.

As a risk practitioner, it is important to determine which critical services should be protected in case of a disaster. The BIA conducted by an organization will support the risk practitioner in recommending a reasonable and appropriate risk response and guide senior management in selecting appropriate mitigation strategies.

A risk assessment is the process of identifying threats to the business operations, the likelihood and impact of the threat materializing, assessing the controls that are already in place, and implementing additional controls that will reduce the risk to an acceptable level.

> **Important note**
>
> More specifically, the primary goal of BIA is to show how quickly an organization should recover critical business operations in case of an incident to avoid further damage; however, the primary goal of a risk assessment is to identify potential threats to the organization and surface the risks that may occur.

In the next section, we will review the concepts related to BIA in detail.

Key concepts related to BIA

Before we move forward with looking at additional concepts in the BIA realm, it would be helpful to differentiate between two major terms that are often incorrectly used interchangeably. These terms are **disaster recovery** (**DR**) and BC.

DR refers to the ability to restore the data and applications that run the business, that is, data centers, servers, or other infrastructure, and how quickly data and applications can be recovered and restored. The speed at which data and applications can be recovered and restored is a crucial aspect of DR. Alternatively, BC planning refers to a strategy that lets a business operate with minimal or no downtime or service outage. It can be said that DR is a subset of BC planning. It is a broader strategy that aims to ensure uninterrupted business operations with minimal to no service disruptions.

It can be inferred that DR is a component of the overall BC planning. An example of DR could be a scenario where an organization's data center is hit by a natural disaster such as a hurricane or earthquake. This could result in damage to the servers and other infrastructure, leading to a loss of critical business data and applications. In response, the company's DR plan would be to restore the data and applications from backups or offsite storage. Similarly, an example of BC could be a scenario where an organization encounters a power outage or unforeseen event, such as COVID-19, that makes it impossible for employees to access critical systems and applications. In response, the company's BC plan would involve procedures for maintaining essential business operations and provide the ability for employees to work remotely as quickly as possible. This could include transferring operations to a secondary site or providing the option to work from home for employees. A BIA serves as the initial point for DR planning for IT systems. A risk practitioner should ensure that sufficient controls and tests have been performed on an application to withstand the various factors determined for it. A completed BIA should help the organization identify the following:

- **Recovery point objective** (**RPO**): Maximum acceptable data loss following an unplanned event
- **Recovery time objective** (**RTO**): Maximum length of time for which a business process can remain unavailable before the business unit's operations are significantly impaired

- **Maximum tolerable downtime (MTD)**: The total amount of time stakeholders are willing to accept for a business process outage or disruption, including all impact considerations

Let's try to understand these terms with the help of an example. Two application owners want to perform BIA on their respective applications. For application A, the application owner confirms that the RPO is four hours – that is, the business will not be impacted if they lose data up to the last 4 hours – the RTO is 8 hours – that is, the business will not be impacted if the application is in downtime for the next eight hours – and the MTD is 12 hours – that is, the business will not be materially impacted if the application is down for the next 12 hours.

Now, the owner of application B performs BIA on that application and reckons that for application B, the RPO is 6 hours, the RTO is 12 hours, and the MTD is 16 hours.

With the help of these metrics, the risk practitioner can determine that application A is more business-critical than application B, and hence in the event of a disaster, the priority for recovery will be application A due to its shorter RTO.

> **Important note**
> The RPOs are always backward-looking and the RTOs are always forward-looking.

The following diagram depicts the disaster planning flow from initiation to RTOs:

Figure 9.1 – Disaster planning flow

In the next section, we will review the difference between inherent risk and residual risk.

Understanding types of risk

Risk and opportunity are two sides of the same coin. As such, it is inevitable that an organization that is looking to expand into new territories and harness the presented opportunities will encounter risks. The goal of a risk practitioner is to manage this risk so that an organization can continue to leverage these opportunities while balancing risk.

There are three types of risks that a risk practitioner should be aware of – inherent risk, residual risk, and current risk. Let us review each of these in detail:

- **Inherent risk**: The level of risk present without considering the actions or controls that will be implemented. This is the risk that is ever-present and is specifically not avoided.

- **Residual risk**: The level of risk after implementing the controls is considered residual risk. Residual risk is calculated by subtracting the effectiveness of control from the inherent risk. *Residual Risk = Inherent Risk – Implemented Controls.*

- **Current risk**: The point-in-time risk for an asset at any given time is considered the current risk. The current risk is prone to changes considering the relative threats and is a more accurate measure of control effectiveness.

Let's look at the relationship between inherent and residual risk with the help of a diagram:

Figure 9.2 – Relationship between inherent and residual risk

You may have noticed that we did not include *current risk* in the preceding diagram. This is because the current risk is always determined at a specific point in time, whereas residual risk is the risk that remains after controls have been implemented. Current risk focuses on the likelihood and potential impact of an adverse event occurring at that moment, whereas residual risk is the risk that remains even after all possible steps have been taken to mitigate it.

To illustrate these concepts, let us look at the following example. Consider a laptop that is used by an employee of an organization. Initially, the laptop did not have an antivirus solution installed and hence the *inherent risk* of it being affected by a variety of viruses was very high. After a few weeks, the IT team rolled out an antivirus solution, which significantly brought down the risk of the laptop being infected by a virus. This remaining risk is *residual risk* and the installation of antivirus is the control. Now, malicious actors release thousands of viruses every day, and therefore the *current risk* of the laptop getting infected by a virus varies on a day-to-day basis.

Summary

At the beginning of this chapter, we learned about the differences between risk assessment and BIA. We learned that the primary goal of BIA is to determine how quickly critical business operations should be recovered in case of an incident to avoid further damage; however, the primary goal of a risk assessment is to identify potential threats to an organization and surface the risks and implement adequate measures.

We then learned about related concepts, such as BC, DR, RPO, RTO, and MTD, which speak to how an organization should determine the recovery objectives of critical systems. In the next section, we switched gears to learn more about inherent risk, residual risk, and current risk, which helps risk managers quantify the remaining risks after all the controls are implemented.

In the next chapter, we will learn about risk response and control ownership, which also marks the beginning of *Domain 3 – Risk Response and Reporting* per the official CRISC exam outline.

Review questions

1. Which of the following statements is *false* regarding risk assessment and BIA?

    A. BIA helps us identify the RPO, RTO, and MTD for critical assets.

    B. Risk assessment helps us identify risk mitigation plans.

    C. A successful business continuity plan needs BIA.

    D. BIA and risk assessment are the same.

2. Which of the following is backward-looking in relation to BIA?

    A. RPO

    B. RTO

    C. MTD

    D. SSO

3. Which of the following helps identify the time for which a critical asset can remain unavailable without significantly impairing the business?

    A. RPO

    B. RTO

    C. MTD

    D. SSO

4. The risk after implementing controls is called ___.

    A. Current risk

    B. Residual risk

    C. Inherent risk

    D. Total risk

5. The risk after implementing controls in a real-time scenario against threats is called ___.

    A. Residual risk

    B. Inherent risk

    C. Current risk

    D. Total risk

Answers

1. **D**. Risk assessment and BIA are not the same.

2. **A**. The **recovery point objective** (**RPO**) helps an organization identify how much data they can lose with minimum impact on services and hence it's backward-looking.

3. **B**. The **recovery time objective** (**RTO**) helps an organization determine the time for which it can be unavailable without impacting the business.

4. **B**. *Residual Risk = Inherent Risk – Controls.*

5. **C**. Current risk is the residual risk in a real-time threat scenario.

Part 4: Risk Response, Reporting, Monitoring, and Ownership

In this part, you will get an overview of the different risk response strategies and the goals of risk management. You will also get an understanding of third-party risk management programs and techniques for managing exceptions. In addition, you will learn about control design and implementation, and lastly, you will receive a primer on log aggregation, risk and control monitoring, and reporting.

This part has the following chapters:

- *Chapter 10, Risk Response and Control Ownership*
- *Chapter 11, Third-Party Risk Management*
- *Chapter 12, Control Design and Implementation*
- *Chapter 13, Log Aggregation, Risk and Control Monitoring, and Reporting*

10

Risk Response and Control Ownership

This chapter marks the beginning of *Domain 3: Risk Response and Reporting* for CRISC. This domain represents 32 percent (approximately 48 questions) of the revised CRISC exam. As a reminder, *Domain 2* of the CRISC exam and the material we learned until *Chapter 9, Business Impact Analysis, and Inherent and Residual Risk*, focused on *IT risk assessment*, which relates to IT risk analysis and assessment. This and the following three chapters focus on risk response, control design and implementation, and risk monitoring and reporting.

The aim of this chapter is to introduce the concepts of **risk response and monitoring** and **risk and control ownership**, take a deeper dive into the **risk response strategies** – mitigate/accept/transfer/ avoid – and ultimately learn about **risk optimization**.

In this chapter, we will cover the following topics:

- Risk response and monitoring

- Risk owners and control owners

- Risk response strategies

- Risk optimization

With that, let us dive into the first section, on risk response and monitoring.

Risk response and monitoring

In the last few chapters, we looked at the best practices for performing risk identification and the best practices for risk assessment. In this and the upcoming chapters, we will learn about the various practices of risk response. The following diagram illustrates the IT risk management life cycle.

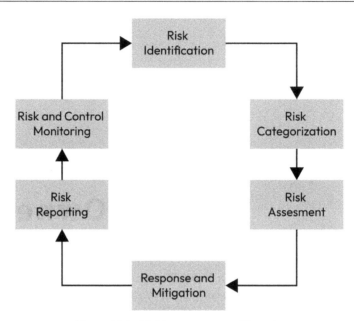

Figure 10.1 – IT risk management life cycle

There can be multiple responses to a risk; however, the job of the risk manager is to assess each of the responses with respect to the budget, time, external regulatory factors, and any disruptions to the current services and identify the response that would most optimize the risk for available resources at the time. The risk manager should then propose these responses to management and relevant stakeholders to obtain buy-in and implement the agreed controls in a reasonable timeframe.

It is important for an organization to monitor the implemented solution over time to confirm that the implemented controls are still relevant to the associated risks. This ensures that the implemented solutions are kept updated with the changing threat landscape and are not outdated.

Let's consider this example to understand the risk response and monitoring in detail. Consider that an organization faces thousands of malicious network connection requests every day from IP addresses all over the world. In response to these attacks, the security analyst would implement a set of firewall rules to thwart malicious traffic and prevent unauthorized access to the network. However, the attack patterns from bad actors are ever-changing, so it is important to monitor these firewall configuration rules and update them regularly to ensure a weakness in the security of the network will be found prior to exploitation.

In the next section, we will learn about the importance of the risk owner and control owner and the difference between each.

Risk owners and control owners

In the previous chapters, we learned about various methods for performing a risk assessment and the importance of having a risk register to catalog all organizational risks in one place. An extremely important part of the risk catalog is having an owner for each of the identified risks to ensure the accountability of these risks is considered and a dedicated individual can be reached to approve the risk response strategy.

In the absence of a **risk owner**, the organization will have a difficult time finding the accountable individual responsible for risk treatment and the risks may go unnoticed. The risk owner should be a manager or a member of the executive committee that is relevant to the identified risks so that they can provide the budget and mandate the risk response based on the risk practitioner's guidance.

Similarly, each risk should have a single risk owner who can speak with authority on the risk response and attest true accountability to the risk. There may be cases where a risk might impact multiple business units and it might not be easier to decide on the owner easily. In these cases, the risk owner should be the leader who can provide the budget, resources, and expertise to remediate the risk. A risk owner owns the loss incurred due to a realized risk scenario.

> **Important note**
> The purpose of risk ownership is to ensure accountability for risk remediation and the risk owner should be recorded in the risk register.

Just like a risk owner, an organization should also have a dedicated **control owner** for each control. In the absence of a control owner, the controls may become outdated and there will be no oversight of their performance with respect to changing risks. A control owner should be the person who implements a control to mitigate risk or is responsible for the oversight of the effectiveness of the control. Unlike risk owners, it is often difficult to determine a control owner at the time of risk assessment. However, the control owner should be noted in the risk register as soon as they are identified.

It is acceptable for the risk owner and the control owner to be the same person, but this should be avoided to maintain the segregation of duties between both roles. This scenario is especially true in smaller organizations where the same role may end up serving multiple responsibilities as they might have the most comprehensive understanding of a given risk and the controls to mitigate it. For example, consider a software development company where a security analyst is designated as the risk owner for potential data breaches. The security analyst identifies this risk, understands its potential impacts on the project, and outlines the necessary controls that need to be implemented. They could decide, for example, that a control required to mitigate this risk is encrypting all sensitive project data.

Because the security analyst best understands the risk and the control, they become the control owner too. They are responsible for ensuring that the team implements and maintains data encryption practices throughout the project. In the next section, we will learn about risk mitigation strategies.

Risk response strategies

Risk response or **risk treatment** is a set of actions that are taken to manage the risk. This is the process of the selection and implementation of measures to optimize risk. The following are the four different ways to respond to a risk:

- **Mitigate**: Risk mitigation is the management of risk through the implementation of countermeasures and controls. The risk practitioner must always keep in mind that the cost of mitigating a risk should be less than the effective risk. The objective of risk mitigation is not to terminate the risk but to bring it down to an acceptable level. The following are a few examples of risk mitigation:

  - Installing anti-malware software to reduce the risk of malware

  - Performing regular backups to reduce the risk of data loss

  - Updating/patching the systems periodically to reduce the risk of running vulnerable software

  - Documenting and testing incident response, business continuity, and contingency plans to ensure the right individuals are informed of such plans when required

- **Acceptance**: Risk acceptance is the decision to accept a risk, made according to the risk appetite and risk tolerance set by senior management where the enterprise can accept the risk and incur the losses. It is important for the risk manager to ensure that the management and relevant stakeholders are fully aware of the consequences of accepting the risk. There should be an absolute *decision* made by the management team on risk ownership. Once the risk is accepted, it should be noted in a system of record and revisited on at least an annual basis to assess the original asset value and the current level of risk and determine whether the current risk response is still acceptable or a new risk response is warranted. The organization may choose to accept a risk within or outside its risk appetite, but should always accept the risk if it falls within the risk tolerance threshold. The following are a few examples of risk acceptance:

  - Not patching a non-critical system

  - Continue using the end of life of software as it's easily replaceable

- **Transfer/sharing**: Risk transfer is the process of assigning risk to another enterprise or partner, usually through the purchase of an insurance policy or by outsourcing the service. Risk transfer is often not the first choice as a risk response and is better suited for risks with very low likelihood and very high impact. The outcome of the risk transfer scenario is often monetary indemnity by the third party, but the loss due to reputational damage still falls on the organization. The following are a few examples of risk transfer:

  - Availing cyber insurance from an insurance provider

  - Sharing security responsibility with **software-as-a-service** (**SaaS**) providers

- **Avoidance**: Risk avoidance is the process of systematically avoiding risk, that is, terminating the risk altogether. This risk response is best suited for scenarios when the risk cannot be addressed by the rest of the response strategies to align with the organization's risk appetite or risk tolerance. As this risk response is often irreversible, the decision to avoid the risk must come from senior management and the pros and cons should be detailed extensively with the risk manager before making the decision. The following are a few examples of risk avoidance:

  - Terminating a project due to resource constraints

  - Shutting down an office location due to natural hazards

While these risk response strategies are optimal, a risk manager's goal is to apply one or a combination of these strategies to optimize the risk. The purpose of risk response is not to eliminate the risk altogether (which is impossible anyway) but to optimize it such that it can also be seen as an opportunity. Let's look at achieving risk optimization in detail in the following section.

Risk optimization

Over the last few chapters, we learned that all risks are not the same and different risks demand different risk responses. That said, the goal of all risk responses is to optimize the risk as much as possible. In some cases, the risk responses are immediately apparent; however, other risks require detailed analysis to provide a response that is best aligned with the organization's goals and business objectives. An organization can choose a risk response based on the following factors:

- Risk category (critical/high/medium/low)

- The cost of associated risks

- The cost of risk response, such as the cost of implementing controls or insurance premium

- The availability of controls

- Available skillsets

- The complexity of implementing controls

- Resources and budgeting

- The alignment of the risk response with organizational strategy

- Compatibility with current controls

- Contractual requirements

- Legal and regulatory requirements

The risk manager must work with senior management to propose a business case for the risk response. The business case should include the **cost-benefit analysis** of adopting a particular risk response strategy, along with the **return-on-investment thesis** so the senior management can make an informed decision on the risk optimization strategy.

Summary

At the beginning of this chapter, we learned about the importance of risk response and monitoring. We then learned about the roles of risk owners and control owners. It is important for the risk manager to be aware of this ownership of risks and controls to take action on them and define the relevant response strategy. The next section then covered the risk response strategies – mitigation, acceptance, transfer/share, and avoid – that a risk manager can use to respond to a risk. We also noted that the goal of risk response is not to remove the risk altogether but to optimize it and use it as an opportunity instead. We then learned about the factors that the risk manager and the management team must consider before proceeding with a risk response that includes a cost-benefit analysis and thorough diligence on return on investment.

In the next chapter, we will learn about **third-party risk management**.

Review questions

1. The *primary* reason for having a risk owner is to:

    A. Leverage their resources

    B. Ensure accountability

    C. Provide flexibility

    D. Seek assistance

2. Which of the following is *not* a risk response strategy?

    A. Accept

    B. Transfer

    C. Treat

    D. Mitigate

3. A risk manager determines that the primary controls for a risk are not sufficient and hence decides to implement a compensating set of controls. This is an example of:

    A. Risk mitigation

    B. Risk acceptance

    C. Risk sharing

    D. Risk avoidance

4. Senior management has decided to buy insurance for an earthquake-prone site. This is an example of:

    A. Risk mitigation

    B. Risk acceptance

    C. Risk sharing

    D. Risk avoidance

5. A risk manager proposes that management terminates onboarding a new tool as it does not adapt itself well to the organization's changing requirements. This is an example of:

    A. Risk mitigation

    B. Risk acceptance

    C. Risk sharing

    D. Risk avoidance

6. An organization decides to allow all employees to work remotely for a brief period of time due to geographical risks. This is an example of:

    A. Risk mitigation

    B. Risk acceptance

    C. Risk sharing

    D. Risk avoidance

Answers

1. **B.** The risk owner for a particular risk drives accountability in remediation.

2. **C.** *Risk response* and *treat* are often used interchangeably but the latter is not a risk response strategy in itself.

3. **A.** Implementing additional controls is an example of risk mitigation.

4. **C.** The senior management has decided to share the risk with the insurance company.

5. **D.** The risk manager has proposed to terminate the onboarding project and hence proposed a risk avoidance response.

6. **B.** The organization allowing employees to work remotely is an example of them accepting the risk of work-from-home arrangements.

Third-Party Risk Management

So far, we've learned about IT risk management and the different methods to perform a risk assessment and response, as well as monitoring. In this chapter, we will dive deep into **third-party risk management** (**TPRM**), how to assess downstream third parties (vendors) and support businesses for upstream third parties (customers), and how to manage emerging risks. We will also look at how to manage issues, findings, and exceptions that may impact the business operations of an organization.

This chapter aims to help you learn about the concepts of TPRM and how to perform an effective third-party risk evaluation. We will also learn about issues, findings, and exceptions and how to manage them effectively.

In this chapter, we will cover the following topics:

- The need for TPRM
- Managing third-party risks
- Upstream and downstream third parties
- Responding to anomalies

With that, let's dive into the first section: *The need for third party risk management.*

The need for TPRM

Before we start learning about TPRM, I think we should talk a bit about why these third parties are required in the first place and what specific purpose they serve for the contracting organization.

Third-party outsourcing is a form of delegating services to another party, such as day-to-day operations, software services, storage, compute, networking, and more, so that the enterprise can focus on its most essential services while delegating the services that can be performed by another organization.

The relationship between the enterprise and the third party is defined in a legally binding contract. The contract includes the set of provisions that the enterprise and hence the outsourcing organization needs to adhere to, such as data storage, compliance with local laws and regulations, jurisdiction in case of disagreements, indemnification clauses, payment terms, **service-level agreements** (**SLAs**), and security and privacy requirements that the third party needs to adhere to. The risk practitioner should play an important role in advising the appropriate security, confidentiality, availability, privacy, and regulatory requirements that the agreement should cover.

Though these contracts are tailored to ascertain that third parties agree to comply with the requirements laid down in the contract, the organization needs to perform its due diligence and due care to ensure that the outsourcing organization can meet the contractual requirements. The outsourcing enterprise needs to understand that regardless of employing a third party to perform certain tasks, the actual ownership of data and business processes remains within their scope and not the third party. This risk can be mitigated by including indemnification clauses that require the third party to repay any losses that cannot be avoided.

A simple example to understand this is when an organization uses a service such as **Amazon Web Services** (**AWS**) or similar for its **Infrastructure-as-a-Service** (**IaaS**) requirements of storage, compute, and networking. Here, AWS is required to provide all these services per the SLAs promised on its website (`https://aws.amazon.com/legal/service-level-agreements`), but the ultimate risk will still lie with the enterprise if AWS is not able to fulfill its SLA obligations. They may compensate for breaching the SLA for providing certain credits or similar but the onus of keeping the services up and running is with the enterprise.

In the next section, we will look at the process that an enterprise can follow before outsourcing to manage third-party risks effectively.

Managing third-party risks

Whenever an organization determines a service that needs to be outsourced, a risk practitioner should be involved in assisting the business in determining the right partners, as well as performing due diligence on the selected vendor. The typical process to determine the right partners and manage the third-party risk should go like this:

1. The business process owner comes up with a use case for outsourcing a service to a third party and has all the necessary approvals from relevant stakeholders.

2. A **request for proposal** (**RFP**) or similar is published or key players in the space are reached out to so that they can assess the availability and alignment of the requirements of the organization.

3. Of all the vendors, a selected few are moved to the next stage so that they can demonstrate how their capabilities are aligned with the requirements of the organization, any niche features that are not available with other vendors, and budget considerations.

4. At this stage, the enterprise should sign a **non-disclosure agreement** (**NDA**) to protect the interests and intellectual property of the enterprise from being disclosed to unauthorized personnel.

5. The risk practitioner should also be involved at this stage and assess the third party on their security and privacy controls. Here are a number of questions that the risk practitioner may find useful in their security assessment:

- Does the vendor follow standard hiring practices? Are adequate background checks such as education, work history, criminal record, and more performed on the prospective candidate and current employees?

- Are the employees trained properly to carry out the assigned tasks?

- Does the vendor have appropriate incident response mechanisms, and would they be able to notify of such an incident in a reasonable time frame?

- Were there any breaches or critical incidents that the organization should be aware of in the recent past? If yes, how did the vendor respond?

- Does the vendor agree to indemnification clauses laid out in the contract?

- Does the vendor follow standard change management practices for the services they provide?

- Does the vendor agree to conduct external audits such as ISO 27001/SOC 2/HITRUST at least on an annual basis and provide a report? Were there any findings that required a **corrective action plan** (**CAP**), and if yes, has the CAP been worked on?

- Does the vendor agree to conduct an annual penetration test and share the results of the test? Were all the open findings closed within a reasonable time frame per severity?

- Where would the vendor store the organization's data? How would the vendor determine who should have access to that data?

- How would the vendor maintain physical and/or logical segregation between data?

- How would the vendor respond to the requests from law enforcement?

- Has the vendor performed a business continuity/disaster recovery test in the past year, and would the vendor be willing to share the results of those tests?

- Has the vendor complied with its SLA obligations in the recent past? If not, was the vendor able to compensate for those breaches appropriately?

- How would the vendor handle the organization's data after the contract has been terminated? Would they be willing to provide a certificate of destruction stating that the data has been returned or destroyed per an industry-standard technique such as NIST 800-88?

- Does the third party comply with federal and state laws such as GDPR or HIPAA? If yes, would they be willing to sign appropriate legal documents such as a **Business Associate Agreement** (**BAA**), **Data Processing Addendum** (**DPA**), and more?

- Does the vendor support a secure transmission protocol such as **Secure File Transfer Protocol** (**SFTP**) for confidential data transfer?

- And many more…

Of course, this is just a set of prompts that the risk practitioner should keep in mind while assessing a third party, but this should be tailored per the services provided by the vendor and the criticality of those services. For instance, a service such as AWS that provides entire infrastructure services and is critical to a business operation should go through a thorough security assessment compared to a service that provides spell-checking software.

This process is also called a **third-party security risk assessment** or **vendor security risk assessment**. Some organizations choose to perform the vendor risk assessment after the proof of concept, whereas others perform it beforehand. Regardless, the important takeaway from this section is to perform this assessment before onboarding a vendor and granting them access to our systems and data.

In the next section, we will look at how we can differentiate between upstream and downstream third parties.

Upstream and downstream third parties

Often, when we think of third parties, we only think about the vendors providing services to us. However, there is another set of third parties that are equally if not more important than the vendors – our customers. I am not sure whether this is a term that is used in the industry to describe customers, but for the sake of this chapter, we will consider downstream third parties as vendors providing services to us and upstream third parties as customers to whom we provide services.

While we assess our vendors and perform due diligence, our customers must perform the same due diligence on us. Therefore, it is important to ensure that the organization maintains a robust internal risk management and cybersecurity program.

One of the best ways to streamline all the components of a risk management program to satisfy third-party requirements is to conduct an external certification such as ISO 27001 or HITRUST CSF or perform an independent audit for a report, such as SOC 2. All these certificates and reports perform the same function – that is, they assess an organization on a defined set of controls per the scope of services and provide an assurance to the customers that the organization has designed and implemented a standard set of controls. From my experience, a SOC 2 report or HITRUST CSF is favored more in the US. Here, the latter is more catered toward healthcare companies, whereas the ISO 27001 certification is more recognized in the European Union/UK.

Though these external audits provide reasonable assurance and guidelines for implementing the necessary controls, the organization should consult with a risk practitioner to ensure that the controls that have been implemented to satisfy the audit requirements will not adversely affect the security of systems or the productivity of developers. Again, this is subjective and can vary for each organization but opting to go for an industry standard and having a baseline to start with always goes a long way compared to starting from scratch.

Responding to anomalies

Regardless of stringent security controls, an organization will always have some issues and exceptions. The goal of a risk practitioner is to ensure that sufficient controls are put in place and procedures are developed in the case of an issue or exception that might pose a risk. For instance, an organization may have implemented an overarching policy of disabling USB access for all employees, but it may be required by the sales team to show a demo of an application or the developers to run a code snippet and perform thorough testing. In those cases, the risk manager should strive to balance such one-off cases by defining a mechanism to manage these requests. In the following section, we will review a few ways to manage these issues, findings, and exceptions.

Managing issues, findings, and exceptions

The following are a few formal approaches to managing issues, findings, and exceptions:

- **Configuration management**: Configuration management requires a baseline/standard set of controls for all the systems of an enterprise. This reduces the complexity by simplifying planning, testing, implementation, and maintenance activities. The risk practitioner should determine that the standard configurations are established, documented, and approved and verify that these are followed by the respective teams. The risk practitioner should also ensure that such configuration baselines are updated as and when required as per the software upgrade or changes in control requirements.

  A good example of configuration management is baselining the laptops of all employees in an organization and hardening them to remove unnecessary services or block the installation of custom software by the end users. The **Center for Information Security** (**CIS**) (`https://www.cisecurity.org`) provides such baselines for almost all general-purpose software services.

- **Release management**: When software is developed, it follows a certain development cycle. It is expected that when substantial changes are ready to be released – that is, they are ready to be moved from development to a production environment – the development team will perform the necessary steps to ensure that the release will be coordinated with the production team, along with additional end user testing. This enables the next version to be released with minimal downtime and fewer errors for the end users.

- **Exception management**: There will be instances when a deviation from the organization's policy and procedure will be required for business reasons. A process must be established and communicated to all employees to ensure that such exceptions can be documented and approved by the right teams. These exceptions often bring an unknown risk to the organization, so the risk practitioner should review and confirm the requirement of such exceptions, at least on an annual basis.

  An example of an exception could be an employee asking for access to print official documents for business reasons. The risk practitioner should document this access to print documents for that employee as an exception and confirm at least annually that such an exception is still required.

- **Change management**: Requests to change systems or configurations should undergo a formal review and approval. To effectively manage such changes, a **change advisory board** (**CAB**) is created that includes representatives from multiple departments, such as IT, security, engineering, and more. In an effective change management process, any changes should be submitted to the CAB, which verifies the following:

  - The requested change will not negatively affect the risk profile or security

  - The change is formally requested, justified, approved, and documented

  - The change is scheduled at a time convenient for business and IT

  - The change will not result in undue impact or major downtime for the end users

  - All relevant stakeholders affected by the change are informed beforehand

  - The requested change has followed implementation guidelines, has gone through robust testing, and has a rollback plan

  - The change will not compromise the security baselines

  The primary goal of such a CAB is to balance the required changes with preserving system reliability and stability. Any changes to the production environment without sufficient adjustments could expose the organization to unidentified risks.

  The risk practitioner should facilitate a different process for emergency changes. All emergency changes should also undergo the same process that we've just discussed and be presented to the CAB to ensure the visibility of such changes to all the relevant stakeholders.

We will review the key points and summarize them in the following section.

Summary

At the beginning of this chapter, we learned about the risk posed by third-party entities and how it can be managed. We then learned about the importance of managing downstream as well as upstream third-party relationships. With the recent trends and an uptick in third-party attack vectors, risk managers should keep themselves abreast of the latest trends and ensure that the risk posed by these threat actors can be minimized by implementing a TPRM program. Next, we learned about issues, findings, and exceptions and the role of configuration, release, exception, and change management to manage these risks. Finally, we learned about the importance of CAB in approving these changes. The goal for risk practitioners is to strike a balance of security and usability without compromising the organization's security goals.

In the next chapter, we will learn about control design and its implementation.

Review questions

1. Which of the following would bind a third party to provide monetary credits to the organization in case of a service failure?

    A. Master service agreement

    B. Service-level agreement

    C. Non-disclosure agreement

    D. External audit

2. Which of the following should be signed with the third party to protect the intellectual property and interests of the organization?

    A. Master service agreement

    B. Service-level agreement

    C. Non-disclosure agreement

    D. External audit

3. Which of the following is *not* the final output of an external audit?

    A. SOC 2 report

    B. ISO 27001 certification

    C. HITRUST certification

    D. Non-disclosure agreement

4. The IT team is implementing new software across the organization and is defining the baseline control settings for end users. This is an example of ____.

    A. Release management

    B. Change management

    C. Configuration management

    D. Exception management

5. The risk practitioner should review and verify the granted exceptions are still required at least ____.

    A. Weekly

    B. Monthly

    C. Quarterly

    D. Annually

6. The CAB should consist of which of the following?

   A. Only IT

   B. IT and engineering

   C. IT, engineering, and security

   D. All relevant stakeholders

Answers

1. **B**. A breach in the SLA allows the organization to demand monetary credit.

2. **C**. Non-disclosure agreements protect the intellectual property and interests of the organization.

3. **D**. Non-disclosure agreements are agreed upon and signed within the organization and by third parties. All the other options are the results of an external audit.

4. **C**. Baseline controls are set as part of configuration management.

5. **D**. The granted exceptions should be verified at least annually.

6. **D**. The CAB should consist of all relevant stakeholders.

12
Control Design and Implementation

As we learned earlier in this book, **risk mitigation** is one of the most common responses in risk management. A risk manager needs to be aware of adequate risk mitigation techniques to reduce the risk to an acceptable level. **Control design and implementation** is one of the most important steps in risk mitigation. With the ever-changing threat landscape, the controls that are implemented today may become irrelevant tomorrow, and therefore, controls should be reviewed periodically to determine and continue their effectiveness.

This chapter aims to help you learn about the different types of controls, standards, frameworks, and methodologies for control design and selection, as well as how to implement them effectively. We will also learn about several control techniques and methods to evaluate them effectively.

In this chapter, we will cover the following topics:

- Control categories
- Control design and selection
- Control implementation
- Control testing and evaluation

With that, let's dive into the first section: *Control categories*.

Control categories

Before we jump right into the control types, I think it is important to learn a bit about what constitutes a **control**. A control is a measure that helps reduce risk and improve the security posture of the organization. This control can be technical, such as antivirus software, something physical, such as a turnstile, or a policy document that dictates the ideal course in business operations.

These controls can be categorized as follows:

- **Preventive** (also known as **preventative**): These controls prevent any security violations and practices. Installing antivirus software to prevent malicious software or a firewall from blocking unknown traffic is an example of preventive control.

- **Detective**: These controls detect violations of security policies and practices. **Intrusion detection systems (IDSs)** or audit logs are examples of detective control.

- **Corrective**: These controls correct a certain issue that has not been prevented or detected and led to an undesired state. Performing a backup and restore test of critical systems is an example of corrective control.

- **Deterrent**: These controls deter malicious actors so that they can't cause any security violations. Installing security cameras in an office facility is an example of deterrent control.

- **Compensating**: These controls compensate for the weakness in other controls or a lack of controls. Rotating the passwords of a shared social media account when an employee is terminated is an example of compensating control.

In the following section, we will learn about the relationship between different control categories.

The relationship between control categories

In the ideal world of a risk practitioner, all malicious actors and threats should be deterred before they cause any damage to the organization; this should act as a first line of defense. If the threat actors are still persistent and carry out the attack, the preventative controls should prevent them from causing any actual damage to the organization and act as a second line of defense.

When the **preventative controls** fail to perform, a set of **compensating controls** should compensate for the lack of controls, and if the issue does become an event or incident, the detective and corrective controls should be able to detect and correct it, respectively. Therefore, a compensating control can be preventative or detective in nature. This is shown diagrammatically in the following figure:

Figure 12.1 – Relationship between control categories

In the next section, we will learn about control design and selection.

Control design and selection

If someone asks you in a rapid-fire round what level of control a risk practitioner should implement, the correct answer is always *optimal*. We have touched on this a few times in previous chapters, stating that a control should always be implemented per the risk posed by the threat and evaluated for effectiveness, efficiency, and cost before it's implemented. There is no reason to implement a control that exceeds the cost of assets.

As we discussed earlier, these controls can be either **proactive** (also known as safeguards), in that they will try to prevent the incident from occurring in the first place, or they can be **reactive**, in that once the incident has happened, these controls will assist in detection and correction. In some cases, the risk practitioner will have the option to choose the type of control to be implemented as per the business requirements; however, regardless of the selected control, the main purpose of implementing control should be to keep the risk at an acceptable level.

In some cases, the risk manager can't implement a control without impacting the business. For example, there might be some applications in an enterprise that are managed only by a single individual, and it will not be possible to have role-based access control or segregation of duties implemented for such applications. Here, the risk manager can propose a compensating control, such as a privileged account for privileged activities, to reduce the probability of mistakes, or a better logging mechanism to detect and correct the errors.

In the next section, we will look at different techniques for implementing these controls.

Control implementation

In the previous chapter, we learned about configuration management, which refers to setting initial baselines for systems and tools so that it becomes easier for the relevant teams to install and manage that software. In this context, configuration management is a *preventative control* that ensures that no unapproved software and services are installed on the user's laptop. The same goes for change management as well, where changes have been made to production system code that need to be tested in a test environment before being rolled out to all the users.

On a related note, it is important to ensure that once these changes have been set, the control is also tested in a non-production environment so that any errors from the test environment do not carry over to the production environment and affect a large set of systems or users in terms of unapproved changes. The ideal way to set up this test environment is to have it reflect the production environment as closely as possible so that a thorough test can be performed. Lastly, a final sign-off from the CAB should be obtained before rolling out the changes.

Now, let's look at some control implementation techniques in detail.

Control implementation techniques

There are three control implementation techniques we can use to roll out or update a system, technology, or version:

- **Parallel changeover**: As its name suggests, in parallel changeover, both old and new systems are operated simultaneously. Parallel changeover confirms the reliability and performance of the new system and ensures that the functionalities of the new system are working as intended before removing the old system.

  As both systems work in parallel, it also provides an opportunity for staff to be trained on the new system while they already have access to the old system. Given both systems are working in parallel, the feasibility of rolling back in case of an issue is also an added benefit of this process.

  Now, you might be wondering, "If there are already so many benefits to using parallel changeover, why should I even look for any other changeover technique?" The answer is budgetary constraints. Not all organizations have the required budget to run both systems in parallel because of the system requirements and ensuring that the data remains consistent in both systems.

- **Phased changeover**: In phased changeover, new systems are implemented in a phased manner – that is, old systems are replaced with new systems gradually. Phased changeover allows us to roll out our new features without them impacting the entire system, thereby reducing the risk of a full system outage.

  Though this is more cost-effective than parallel changeover, as it doesn't require old and new systems to run in parallel, the additional complexity in phased changeover is to maintain separate resources and software and hardware environments for both systems.

- **Abrupt changeover**: In abrupt changeover, the old systems are replaced with new systems instantly – that is, the old systems are taken offline as soon as the new systems are implemented. Abrupt changeover is suitable for situations when rolling back is easy and there is no major loss to business processes.

  It is important to note that this changeover is the riskiest of all and should not be performed when the changes affect critical business processes.

The following table aims to summarize these techniques:

Type	Definition	Advantages	Disadvantages	When to Use
Parallel	Both systems run in parallel	Safe, easy to roll back, little to no downtime, and provides the opportunity to train	Expensive to maintain and data consistency challenges	Complex system migration or when the cost of an outage exceeds the cost of parallel systems
Phased	Phased update of the old system to a new one	Cost-effective, reduced risk of system outages, and modular rollback	Resources to maintain separate environments	When working with complex systems with independent modules
Abrupt	Instant update from the old system to the new one	Fast and cost-efficient compared to the other approaches	Very risky, could lead to tremendous business loss, and a complete rollback is required	Minor changes that don't impact critical business processes

Table 12.1 – Control implementation techniques

In the next section, we will learn about the mechanisms to review the implemented controls.

Post-implementation reviews

Post-implementation reviews allow system owners and risk practitioners to review the changes that have been made and learn the key lessons, regardless of whether the implementation is successful or not. A timely post-implementation with all the relevant stakeholders provides an open forum to raise concerns about the steps that did not go well but also accounts for the positives in the entire implementation process. Lessons learned from the post-implementation review should be documented and disseminated to relevant participants as well as a wider audience so that they can also benefit and improve their respective implementation processes.

The following are some questions that might be useful for the post-implementation review committee:

- Did the implemented project meet the business objectives it was initially intended for?

- Did the project satisfy the user requirements that were gathered initially?

- Has the project been completed on time and within the allocated budget?

- Did the team follow a documented **software development life cycle** (SDLC) and was sufficient manual, automated, and security testing performed?

- Have logical and business controls been appropriately defined and implemented?

- What went well and what could be improved?

- How long did it take to fix any issues/bugs that came up during the implementation process?

- Were sufficient resources allocated to complete the project on time?

- Do we have dedicated resources for continuous support?

The post-implementation review document should be considered a live document and updated as and when new learnings come up during the life of the project.

In the next section, we will look into the best practices for control testing and evaluation.

Control testing and evaluation

Testing the effectiveness and efficacy of a control is as important as implementing them. A **risk practitioner** should ensure that implemented controls are tested and evaluated periodically to ensure that they are still relevant and advise the risk owner in case of any gaps that have occurred since the initial implementation. The responsibility to determine the efficacy of controls periodically relies on the **control owner**. Control testing can be either progressive or regressive. **Progressive testing** begins with the requirements and looks for flaws, whereas **regressive testing** works backward from the expectations of the results and known issues to identify causes.

The following are some of the best practices for effectively evaluating controls:

- Never use production data for testing purposes and always produce synthetic data that's as similar to the production data as possible for testing. If there is an absolute need to use the production data in a test environment, then mask sensitive fields, such as user credentials, to prevent confidential data from being disclosed by unauthorized users.

- Keep the test and production environments entirely separate to prevent unauthorized changes in the production environment.

- Maintain version control to track the current iteration and obtain adequate approvals before the code merge, as well as the feasibility to roll back to an earlier version.

- Implement code freeze before pushing the code to production to reduce the likelihood of untested code in the production environment.

- Use access control to determine who can push the test code to production.

- Perform unit, system, integration, performance, stress, and functional tests to confirm that the business case and user requirements will be satisfied with the implementation.

- Perform code reviews to uncover any hardcoded secrets and logical errors.

The risk manager should evaluate the environment and bring any additional controls that should be tested and evaluated. We will now review the key points we learned in this chapter and summarize them.

Summary

At the beginning of this chapter, we learned about the five control categories (preventative, detective, corrective, deterrent, and compensating) and their relationship with incidents. We then learned about how a risk manager should design and select the controls as per the requirements of the business. Next, we learned about the different methods of control implementation (parallel, phased, and abrupt) and the importance of post-implementation review. Finally, we reviewed the best practices for control testing and evaluation.

In the next chapter, we will learn about log aggregation, risk and control monitoring, and reporting.

Review questions

1. Which of the following is *not* a control category?

    A. Additive

    B. Preventative

    C. Deterrent

    D. Detective

2. Implementing controls proactively based on a previous root cause analysis is an example of which control category?

    A. Preventive

    B. Detective

    C. Corrective

    D. Compensating

3. Reviewing audit logs in a SIEM tool is an example of __.

    A. Corrective control

    B. Detective control

    C. Compensating control

    D. Deterrent control

4. Installing security cameras to secure a data center location is an example of __.

    A. Corrective control

    B. Detective control

    C. Compensating control

    D. Deterrent control

5. A risk manager introduced additional oversight of an accounting system where logical controls are not yet implemented. This is an example of ____.

    A. Corrective control

    B. Detective control

    C. Compensating control

    D. Deterrent control

6. An IT system failed while new changes were being pushed to it from the production team and lost some data. The risk manager has advised the IT team to restore the data from previous backups. This is an example of ____.

    A. Corrective control

    B. Detective control

    C. Compensating control

    D. Deterrent control

7. Which of the following control categories are used after the incident has happened?

    A. Corrective, compensating

    B. Corrective, deterrent

    C. Corrective, detective

    D. Deterrent, preventive

8. A risk manager is overseeing a major upgrade to an IT system. The system is business-critical and should not have any downtime. What should be the risk manager's recommendation for the implementation?

    A. Parallel

    B. Phased

    C. Abrupt

    D. Any of the above

9. A system is due to upgrade per module and no two modules are dependent on each other. What would be the most efficient manner to upgrade the system?

    A. Parallel

    B. Phased

    C. Abrupt

    D. Any of the above

10. Who should be the target audience for a post-implementation review?

    A. Developers

    B. Risk manager

    C. Risk owner

    D. All major stakeholders

Answers

1. **A**. Additive is not a control category; all the other options are valid control categories.

2. **A**. Implementing controls proactively is an example of preventive control.

3. **B**. Reviewing audit logs is an example of detective control as the incident has already happened and the goal is to find what caused the incident.

4. **D**. Installing security cameras is an example of deterrent control.

5. **C**. Implementing additional controls instead of the primary control is an example of compensating control.

6. **A**. Performing a restore exercise is an example of corrective control.

7. **C**. Detective and corrective controls are used after the incident has happened.

8. **A**. Since the system is business-critical and should not have any downtime, the risk manager should recommend parallel changeover, where both the old and new systems will run in parallel.

9. **B**. Performing a phased changeover would be the most efficient as no two modules are dependent on each other and the update is modular.

10. **D**. All major stakeholders should participate in the post-implementation review to share their learnings.

13
Log Aggregation, Risk and Control Monitoring, and Reporting

This is the last chapter of *Domain 3: Risk Response and Reporting* and is divided into two parts. In the first part, we will look at the different sources for collecting logs, tools, and best practices to aggregate them, and how to analyze those logs. In the second part, we will look at risk and control monitoring, different control assessments, risk and control reporting methods, different key indicators for an executive summary, and the appropriate audience for each.

The aim of this chapter is to learn about the different methods of log sources, aggregation, and analysis. We will also learn about risk and control monitoring, reporting, and how to present reports effectively.

In this chapter, we will cover the following topics:

- Log aggregation and analysis
- Security information and event management
- Risk and control monitoring
- Risk and control reporting
- Key indicators

With that, let us dive into the first section on log aggregation and analysis.

Log aggregation and analysis

As we learned in the previous chapter, logs play an important role in implementing a detective and corrective control strategy. There are millions and millions of events happening in an organization, and without a proper mechanism to aggregate and analyze these logs, the security team could miss many important events, which could lead to an incident. To understand this, let's consider the example of Google's Gmail. Imagine the number of people inserting an incorrect password for their Gmail account at any given time. A human could enter an incorrect password maybe five to seven times in a minute. But what if Gmail encountered that the same person was trying out different passwords 100 times per minute or maybe 1,000 times per minute? This would be impossible for a human, and that's where this would trigger an alert for Google's **Security Operations Center** (**SOC**). Now, as a SOC analyst, you could review one or two of those attempts at a time. But how would you review hundreds or thousands of these alerts at the same time? This is where log aggregation, analysis, and alerting come in handy. In this case, the log aggregation tool would collect the logs, perform an analysis, and alert the SOC analyst of this malicious traffic. Of course, there are a lot of caveats in the preceding example, such as this: *What if it's just the Google team checking a new defensive strategy they implemented?* In the following sections, we will learn more about log sources, aggregation, and correlation.

Log sources

A risk practitioner has multiple data sources for **logging** and **monitoring**, but the important question that they need to ask is this: *Are all those sources necessary for logging?* There are application logs, network logs, system logs, **threat intelligence** (**TI**) logs, user event logs, and so on, and all of these could make their way to the logging tool. The risk practitioner needs to be mindful that with each additional system that is accounted for logging, the immense volume of overall data collected for logging should not defeat its purpose of identifying events. The risk practitioner should always consider and ask whether this log source plays an important role in the organization's infrastructure and security. If the answer is yes, then it should absolutely be collected and aggregated, or else a local copy of logs should be sufficient. For each log source, the risk manager should aim to record the following minimum information:

- Successful login and logout information

- Failed login and logout information

- Changes to user roles

- System start and shutdown time

- Failure events

- Timestamp for each event

- User identifiers performing configuration changes

- Unique identifiers such as IP address

In the next section, we will look at the importance of log aggregation.

Log aggregation

The logs collected from different sources are well utilized when aggregated collectively in a centralized log management system. **Log aggregation** makes it easier for end users to search, group, and analyze logs to identify patterns and events.

As different systems aggregate, the risk manager should emphasize unifying the formats of different logs for correlation. All the aggregated logs should have the same time format, as facilitated by the **Network Time Protocol** (**NTP**) for clock synchronization.

In the next section, we will learn about how this all comes together in the form of a **security information and event management** (**SIEM**) tool.

SIEM

SIEM systems are integrated data correlation tools that help in integrating different systems and collecting, analyzing, and alerting based on intelligent thresholds. SIEM systems can be used to detect malicious events based on pre-defined signatures or behavior. SIEM systems allow risk practitioners to identify risk and bring it to the attention of management before it materializes by correlating it with similar events.

An important consideration for implementing a SIEM system is the regular monitoring and fine-tuning of alerting rules to reduce false-positive alerts as much as possible. This constant fine-tuning ensures that the risk practitioner is only focusing on important alerts that require action.

A SIEM system can translate all the logs in management reports and dashboards that are important for management reporting, can support compliance requirements for log retention, and will be helpful for forensic analysis. The risk manager should ensure that the logs are stored in an immutable database to ensure the integrity of data aggregated from various systems and maintain read-only access for a majority of users.

A few examples of SIEM tools are Splunk, IBM QRadar, LogRhythm, Rapid7 InsightOps, Alert Logic, and Elastic Security.

In the next section, we will learn about risk and control monitoring.

Risk and control monitoring

The risks to an organization are ever-changing, and so is the risk profile. Risks encountered a year before may not be relevant anymore, and the controls recently implemented for the latest risk may have already become outdated. A risk practitioner should continuously monitor, benchmark, and improve the control environment to meet organizational objectives. The monitoring of controls can be done through self-assessments or independent third-party audits. Exceptions to controls should be reported, followed up, and addressed with corrective actions. In the following section, we will review some techniques that the risk practitioner can implement with the help of risk owners for effective risk and control monitoring.

Types of control assessments

Before we jump into the techniques for **control assessment**, let's briefly review what this term means. Control assessment is the process of evaluating and examining the effectiveness and adequacy of internal controls within an organization. Internal controls are policies, procedures, and practices put in place to mitigate risks, ensure compliance with regulations, safeguard assets, and achieve organizational objectives.

During a control assessment, an organization assesses whether its internal controls are designed and operating effectively to address identified risks. The assessment involves reviewing control activities, such as **segregation of duties (SoD)**, access controls, documentation, and monitoring procedures.

The following control assessment techniques can be used by the risk practitioner to verify the effectiveness of controls:

- **Self-assessments**: These are also known as internal risk assessments or **management self-identified issues (MSIIs)**. The intent of self-assessments is to engage the key resources involved in the regular working of the system and facilitate workshops to identify issues that they often come across but are overlooked. It is imperative to hold these workshops without judgment and finger-pointing to lead an open discussion and come up with risks, corresponding implemented controls, and additional countermeasures that can be implemented to reduce the risks to an acceptable level. This can either be performed by adopting an industry standard, such as the **National Institute of Standards and Technology (NIST)**, **Control Objectives for Information and Related Technology (COBIT)**, or the **International Organization for Standardization (ISO)** *27001* control framework, or by brainstorming for risks that are relevant to the organization.

- **Internal information system (IS) audit**: An internal IS audit is a good way to determine the effectiveness of controls. The risk manager can collaborate with the auditor to provide evidence, and recommendations provided by the auditor can be objectively presented to the management team for control enhancements.

- **Vulnerability assessment**: A **vulnerability** is a weakness in the design, implementation, operation, or internal control of a process that could expose the system to adverse threats, and vulnerability assessment is the process of identifying such weaknesses. Various tools in the industry can help in identifying existing vulnerabilities such as tool misconfiguration or missing updates. The risk practitioner performing vulnerability assessments should have a working knowledge of both technical and non-technical security controls of the system. A major drawback of a vulnerability assessment is it only identifies vulnerabilities that are already known but not yet discovered.

- **Penetration testing**: A penetration test simulates a real attack and identifies vulnerabilities such as business logic errors and cascading vulnerabilities that have not yet been discovered. The tests performed under a controlled environment with the knowledge and approval of the organization are called **white hat testing**. There are three types of such penetration tests—**white box**, **black box**, and **gray box**, each referring to the amount of information provided to the testers. Let's look at what they entail:

  - **White-box testing**: All the required information is provided for testing

  - **Gray-box testing**: Some information is provided, such as user credentials

  - **Black-box testing**: No information is provided to the tester, hence the name

- **Third-party assurance**: An organization can choose to undergo a third-party attestation whereby an independent third-party (external auditor) will review the controls based on a pre-defined standard such as ISO *27001*, the **Payment Card Industry Data Security Standard (PCI DSS)**, the **Health Information Trust Alliance Common Security Framework (HITRUST CSF)**, or the **Statement for Standards for Attestation Engagements (SSAE)** *18* from the **American Institute Certified Public Accountants (AICPA)** **Service Organization Control (SOC)** *2*, SOC *3*. They, they present the organization with a certificate or attestation report as applicable along with the findings. Typically, a third-party assurance carries more weight in building stakeholder trust and confidence due to its independence and robust testing practices.

In the next section, we will look at how these monitored risks can be best reported to management.

Risk and control reporting

In the previous section, we reviewed the importance of risk monitoring and how it can impact an organization's resilience toward malicious attacks. In this section, we will review how those monitored risks and metrics can be best reported to the management team. Different organizations choose different mechanisms to report on risks and controls. Some are okay with sending a brief executive summary, while others need to elaborate with reports and dashboards. There are no right or wrong ways to present these risks to senior management; however, the risk practitioner, as well as the business owner, should tailor the reports and reporting mechanism per the audience. Would it really make any sense to report the number of phishing attempts in the past month to the head of physical security?

Here are some key aspects the risk practitioner should keep in mind while reporting:

- **Audience**: Who is the right audience for the report?

- **Actionability**: Are the metrics presented in the report actionable?

- **Format**: Am I aware of the audience's preferred format?

- **Succinctness**: Is the report succinct and does it show key information relevant to the audience?

- **Source**: Am I confident about the integrity of the data source?

- **Tailoring**: Can I customize the report so that it is most relevant to the primary stakeholders?

- **Timeframe**: Am I presenting a relevant timeframe?

- **Inferences**: Can the audience infer key bullets from this report?

- **Cadence**: Have I agreed on the cadence for sending this report?

This list is in no way exhaustive, and the reporting team should keep this in mind while preparing any management summaries.

There are four primary formats for risk reporting, as follows:

- **Executive summary**: These are effective ways of summarizing key risks and controls in a page or two to provide concise information to the relevant stakeholders. This doesn't need to be in a specific format and is often sent in the body of the email or as an email attachment. This is useful when reporting on an effective project or milestone with some quantified metrics to showcase the effectiveness.

- **Heat maps**: Heat maps are graphical representations in a divided square plot. The plot is usually divided into a 2×2, 3×3, 4×4, and so on dimension and is color coded from the minimum (starting at *0,0*) to the maximum (ending at *n×n*). The *y* axis of the plot typically shows the *impact*, and the *x* axis shows the *likelihood* of risk occurrence. Heat maps are often qualitative, which leaves room for error in judgment at the receiver end.

- **Scorecards**: Risk scorecards simplify risk reporting by aggregating performance and assigning grades to each area. As with heat maps, scorecards are qualitative in nature, which limits their full scope.

- **Dashboards**: Dashboards are a collection of a variety of metrics and indicators in a presentation format. Unlike the previous methods of risk reporting that are driven primarily by qualitative analysis, the aim of dashboards is to present indicators as a combination of qualitative and quantitative metrics. These dashboards are often created on a recurring cadence to showcase trend identification, analysis, and anomalies. Dashboards are often the preferred method for risk reporting due to their flexibility of content (quantitative and qualitative) and the ability to include a number of metrics.

In the next section, we will learn about the metrics that are reported in dashboards and how to select them effectively.

Key indicators

The metrics reported in dashboards are called **key indicators**. Three types of indicators are reported in a dashboard:

- **Key performance indicators** (**KPIs**): KPIs are used to understand and enable the measurement of control performance. The level of performance can be different for each organization, and therefore, the risk manager should strive to define KPIs that make the most sense to the organization's objectives and risk appetite. An example of a KPI could be a reduction in phishing emails after implementing a new tool.

- **Key risk indicators** (**KRIs**): KRIs are considered to be highly probable indicators designed to predict risks that could breach the defined thresholds. The goal of defining KRIs is to monitor and analyze trends, determine the effectiveness and efficiency of controls to make informed decisions for current controls and planned countermeasures, and alert relevant stakeholders when the risk breaches the predefined thresholds. An example of a KRI could be a group of employees not trained in security awareness and continuing to fall for phishing attacks.

- **Key control indicators** (**KCIs**): KCIs are a measure of the effectiveness of controls to indicate a weakness that may increase the probability of risk events. The goal of KCIs is to track the performance of control actions relative to tolerance and provide insight into the effectiveness of controls to keep the risk within acceptable levels. An example of a KCI could be the lack of implemented controls to block phishing emails.

Let's look at how we can select these indicators.

Selecting key indicators

Selecting key indicators is an important step as it defines the entire strategy and often the actions and resources of the management team. A key indicator should be chosen based on **SMART** metrics, defined as follows:

- **Specific** (**S**): Clearly understandable and concise
- **Measurable** (**M**): Can be measured and quantified
- **Attainable** (**A**): Realistic and based on goals
- **Relevant** (**R**): Relevant to a specific activity or goal
- **Timely** (**T**): Timebound and not open-ended

Now, let's briefly summarize what we've learned in this chapter.

Summary

At the beginning of this chapter, we learned about log aggregation and analysis, risk and control monitoring, reporting, and key indicators. We then learned about the importance of effectively collecting and analyzing logs, implementing SIEM systems, continuously monitoring and improving control environments, and tailoring risk reports to the audience. In the following section, we reviewed the importance of various key indicators, such as KPIs, KRIs, and KCIs, for measuring control performance, predicting risks, and assessing control effectiveness. In the final section, we gained insights into managing risks and reporting them in a meaningful and actionable manner.

In the next chapter, we will delve deeper into enterprise architecture and information technology.

Review questions

1. What is the primary purpose of log aggregation in an organization?

    A. To identify and analyze important events and potential incidents

    B. To synchronize the clocks of different systems for accurate logging

    C. To eliminate the need for logging across multiple systems

    D. To provide a centralized database for log storage

2. Which of the following is a commonly used tool for log aggregation and analysis?

    A. SIEM system

    B. Vulnerability assessment tool

    C. Penetration testing tool

    D. Heat map generator

3. What is the role of KRIs in risk monitoring?

    A. To predict risks that may breach predefined thresholds

    B. To measure the effectiveness of controls

    C. To indicate weaknesses in control implementation

    D. To assess control performance against goals

4. Which format of risk reporting is known for its flexibility in combining qualitative and quantitative metrics?

   A. Executive summary

   B. Heat map

   C. Scorecard

   D. Dashboard

5. What are the key characteristics of a SMART metric for selecting key indicators?

   A. Specific, Measurable, Attainable, Relevant, Timely

   B. Standardized, Meaningful, Achievable, Realistic, Timeless

   C. Secure, Measurable, Aligned, Reliable, Trustworthy

   D. Significant, Measurable, Accurate, Relevant, Timed

6. In risk and control monitoring, which technique involves simulating a real attack to identify vulnerabilities that have not yet been discovered?

   A. Self-assessment

   B. Internal information system audit

   C. Vulnerability assessment

   D. Penetration testing

Answers

1. **A.** Log aggregation helps in collecting and analyzing logs to identify significant events and potential incidents that could impact the organization's security.

2. **A.** SIEM systems are commonly used for log aggregation and analysis, providing a centralized platform to collect, analyze, and correlate logs from various sources.

3. **A.** KRIs are designed to serve as highly probable indicators that predict risks breaching predefined thresholds, helping in proactive risk management.

4. **D.** Dashboards are known for their flexibility in combining qualitative and quantitative metrics, providing a comprehensive and visual representation of key risk and control information.

5. **A.** SMART metrics are essential characteristics for selecting key indicators. They should be specific, measurable, attainable, relevant to goals, and time-bound.

6. **D.** Penetration testing involves simulating real attacks to identify vulnerabilities that have not yet been discovered, providing valuable insights into an organization's security weaknesses.

Part 5:
Information Technology, Security, and Privacy

In this part, you will get an understanding of information technology architecture and how it relates to enterprise architecture, and the importance of enterprise resiliency and data life cycle management. In addition, you will learn about the software development life cycle, emerging technologies and their security implications, and, lastly, the fundamentals of information security and privacy principles.

This part has the following chapters:

- *Chapter 14, Enterprise Architecture and Information Technology*
- *Chapter 15, Enterprise Resiliency and Data Life Cycle Management*
- *Chapter 16, The System Development Life Cycle and Emerging Technologies*
- *Chapter 17, Information Security and Privacy Principles*

14
Enterprise Architecture and Information Technology

This chapter marks the beginning of *Domain 4: Information Technology and Security* for CRISC. This domain represents 22 percent (approximately 33 questions) of the revised CRISC exam. These topics build the foundation of an organization and information technology and are essential to learn and understand not only for the exam but also for building a career in the information security domain. In addition, we will be talking about **information technology**, **information security principles**, and **data privacy** in the following chapters.

The aim of this chapter is to introduce the concept of **Enterprise Architecture (EA)**, the **Capability Maturity Model (CMM)**, and **IT operations** such as network and technology concepts. Without a thorough understanding of the following topics, it is difficult to rationalize the security controls that should be implemented to secure IT assets including networks, networking devices, firewalls, and cloud resources effectively.

In this chapter, we will cover the following topics:

- Enterprise architecture
- The CMM framework
- Computer networks
- Networking devices
- Firewalls
- Intrusion detection and prevention systems
- The Domain Name System
- Wireless networks

- Virtual private networks
- Cloud computing

With that, let us dive into the first section, on EA.

Enterprise architecture

EA is the foundation for running any successful business effectively. A business has many parts, such as people, processes, technology, and data, that work together to produce value for customers, with the goal being customers buying from the business and making it profitable. The purpose of EA is to ensure these parts continue to work together and produce value for the business.

Though the scope of CRISC is limited to technology, it is important to understand the four major domains of EA and then drill into the technology architecture:

- **Business architecture**: Business architecture captures how a business operates and defines business processes in the context of the organization. Business architecture defines the role of underlying software architecture to ensure that it doesn't become obsolete with respect to customer requirements.

- **Application architecture**: Applications support businesses. Application architecture defines the software solutions that enable businesses to achieve their objectives and adapt to application changes so they can support changing business requirements.

- **Data architecture**: Applications need data to work. As an organization gathers more and more data, it needs a better way to manage and store that data. Data architecture defines the model for working with different sources and formats of data that applications can use to derive meaningful inferences.

- **Technology architecture**: All three preceding architectures are supported by physical entities such as wires, networking devices, storage systems, hardware components, and software. The technology architecture describes the underlying infrastructure needed to run the business applications to achieve business objectives.

The following diagram shows EA in a nutshell:

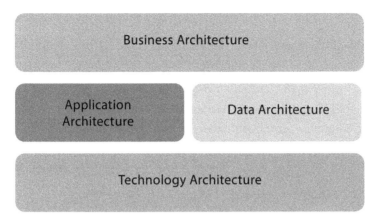

Figure 14.1 – Enterprise architecture

Technology architecture shows the current state of IT, establishes a vision for a future state, and assists the organization in moving from the current to a future state. The goal of a risk manager is to ensure that this move happens with minimal disruption to IT and hence the business. Since the underlying IT systems are dependent on each other, a security issue in one system can affect others; therefore, a thorough assessment of all system components is important to maintain the security of the system.

The Open Group Architecture Framework (TOGAF), the **Zachman Framework**, the **Department of Defense Architecture Framework (DODAF)**, the **Federal Enterprise Architecture Framework (FEAF)**, and **Sherwood Applied Business Security Architecture (SABSA)** are a few examples of commonly used EA.

In the next section, we will learn about the related concept of the CMM.

The CMM framework

The CMM framework provides a structured approach for assessing and improving the maturity and capability of an organization's processes. It was developed in 1986 based on a study of data collected from organizations working with the US Department of Defense.

The term **maturity** refers to the level of formality and optimization of processes, including ad hoc practices, formally defined steps, managed result metrics, and active optimization of these processes. While the CMM was originally developed to improve and objectively assess the ability of government contractors' software development processes, it can also be applied to other processes. Later, in 2006, the Software Engineering Institute at Carnegie Mellon University enhanced the CMM and developed the **Capability Maturity Model Integration (CMMI)**, which has largely replaced the CMM and addresses its limitations.

The following diagram depicts the five levels of the CMM:

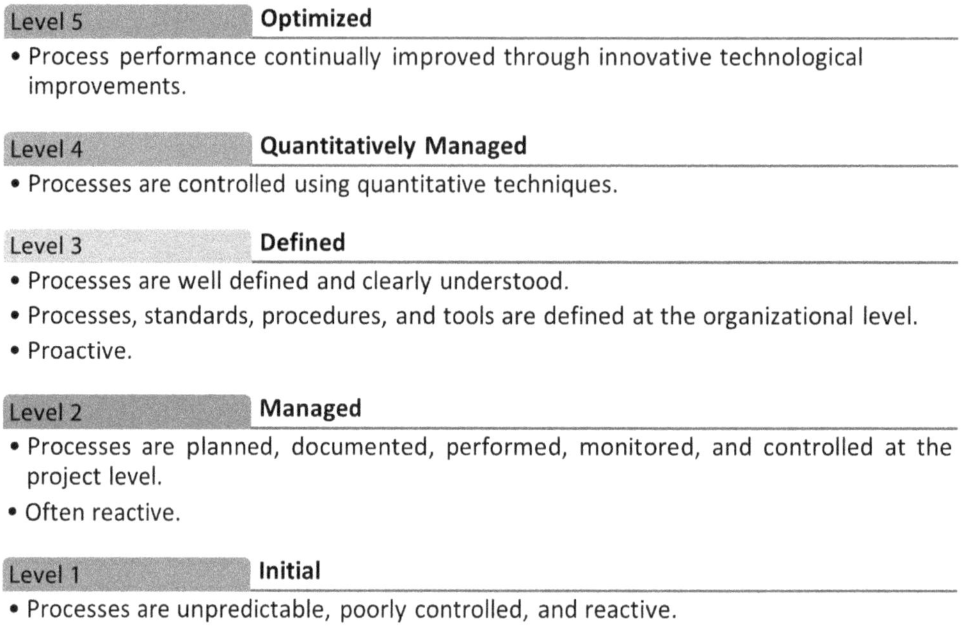

Level 5	Optimized

- Process performance continually improved through innovative technological improvements.

Level 4	Quantitatively Managed

- Processes are controlled using quantitative techniques.

Level 3	Defined

- Processes are well defined and clearly understood.
- Processes, standards, procedures, and tools are defined at the organizational level.
- Proactive.

Level 2	Managed

- Processes are planned, documented, performed, monitored, and controlled at the project level.
- Often reactive.

Level 1	Initial

- Processes are unpredictable, poorly controlled, and reactive.

Figure 14.2 – CMM maturity levels

For a detailed explanation of these levels, please refer to `https://www.isaca.org/resources/isaca-journal/issues/2021/volume-6/building-a-maturity-model-for-cobit-2019-based-on-cmmi`.

In the next section, we will learn about the important components of technology architecture, starting with computer networks.

Computer networks

The term **computer networks** refers to interconnected computing devices that can exchange data and resources. Networks are used for almost all modern technology operations. The following list includes some of the different ways in which they can be used:

- Enabling access to the internet and other applications via a web browser

- Transferring data between individuals

- Transferring data between applications

- Performing data backups

- Controlling remote equipment

- Enabling communication

Just like humans need a common language to communicate with each other, networks need a common protocol to communicate with other networking devices. There are two main models to facilitate this communication:

- **TCP/IP model**: **Transmission Control Protocol/Internet Protocol** (**TCP/IP**) is a set of communication protocols used to connect devices to the internet. TCP/IP is a layered protocol, with each layer responsible for specific functions in the communication process. The TCP/IP model has five layers:

  - The **physical layer** (Layer 1) depicts the physical infrastructure that carries transmissions.

  - The **data link layer** (Layer 2) handles the transfer of data across network media.

  - The **network layer** (Layer 3) moves messages from one network to another using the IP protocol.

  - The **transport layer** (Layer 4) works closely with the IP protocol and performs error checking. There are two protocols on this layer – **TCP** and **User Datagram Protocol** (**UDP**). For the scope of CRISC, only remember that TCP is connection-oriented and guarantees the delivery of data; however, UDP is a connection-less protocol that utilizes its best effort to deliver the data. An example of TCP is email, whereas an example of UDP is Voice over Internet Protocol services such as Skype or WhatsApp calls.

  - The **application layer** (Layer 5) includes processes that use transport layer protocols to deliver data.

- **The OSI model**: Unlike the TCP/IP model, the **Open Systems Interconnection** (**OSI**) model has seven layers:

  - The **physical layer** (Layer 1) is responsible for maintaining a physical connection between devices

  - The **data link layer** (Layer 2) is responsible for the host-to-host delivery of messages

  - The **network layer** (Layer 3) is responsible for the transmission of data from one host to another in different networks

  - The **transport layer** (Layer 4) provides services to the application layer and consumes services from the network layer

  - The **session layer** (Layer 5) is responsible for the establishment of connection, session, authentication, and security

  - The **presentation layer** (Layer 6) extracts the data from the application layer and translates it into a machine-readable format

  - The **application layer** (Layer 7) is also known as the desktop layer and provides an interface for the user to interact with the network

> **Important note**
> The primary difference between TCP/IP and the OSI model is that the former determines the specific requirements for connecting a given computer to the internet, while the latter is a conceptual framework that describes the way network communication occurs in systems that are designed for interconnection.

Now that you know the types of networks, we will learn about devices that enable networking in the next section.

Networking devices

Networking devices are physical equipment that allow hardware devices to communicate and interact with other computer networks. The following is a brief summary of different types of networking devices:

- **Repeater**: A repeater is a two-port (input port and output port) device that regenerates the signal over a network before it becomes weak or gets damaged.

- **Bridge**: A bridge is also a two-port device that joins two networks.

- **Switch**: A switch is a multi-port bridge that provides dedicated pathways based on MAC address association.

- **Router**: A router connects multiple switches and networks to form a larger network by identifying IP addresses.

- **Gateway**: A gateway connects different protocols and networks.

The following table details the functioning of each networking device and its corresponding OSI layer:

Networking Device	Operates on (OSI Model Layer)	Additional Features
Repeater	Layer 1 (physical)	Operates within the network
Bridge	Layer 2 (data link)	Operates between two networks
Switch	Layer 2 (data link)	Performs error checking before forwarding
Router	Layer 3 (network)	Connects two independent networks
Gateway	Layer 3 (network)	Ability to work on different networks

Table 14.1 – Networking device and corresponding OSI layer

In the next section, we will learn about firewalls that can secure these networks.

Firewalls

A **firewall** is a network security device that monitors both incoming and outgoing network traffic and either permits or prohibits specific traffic based on predefined security rules. A firewall can be a physical device (hardware) or software. The following is a summary of different types of firewalls that can be installed as hardware or software:

- **Packet filtering firewall**: These firewalls compare each packet to established criteria and allow/ deny packet traffic based on predefined rules. These firewalls work on Layer 3 (the network layer) of the OSI model.

- **Circuit-level gateway firewall**: These firewalls monitor TCP handshakes and established sessions to determine the legitimacy of the traffic. These firewalls work on Layer 5 (the session layer) of the OSI model.

- **Application-level gateway firewall**: These firewalls provide the most secure connectivity as they examine each layer of communication. These firewalls work on the application layer (Layer 7) of the OSI model.

- **Stateful inspection firewall**: These firewalls keep track of the connections in a table, which allows it to enforce rules based on packets in the context of the communication session. These firewalls work on Layer 3 (the network layer) of the OSI model.

- **Next-generation firewall**: These firewalls combine packet and stateful inspection and prevent intrusions by inspecting connections through deep packet inspection. These firewalls work on Layer 7 (the application layer) of the OSI model.

A risk manager should ensure that firewall rules are documented and reviewed on a periodic basis. It is also important that the employees working on company firewalls are sufficiently trained so they can manage the firewalls effectively and review any suspicious activity.

In the next section, we will look at intrusion detection and prevention systems, which are required when malicious traffic is not blocked by firewalls.

Intrusion detection and prevention systems

The purpose of firewalls is to allow legitimate traffic and block malicious traffic. However, an intrusion system is required in the event that malicious traffic is not blocked by the firewalls. There are two forms of intrusion systems:

- **Intrusion Detection System** (**IDS**): An IDS detects potential malicious traffic but doesn't block the traffic. Whenever an IDS detects malicious traffic, it sends an alert to the respective teams to investigate the alert. Therefore, it's critical to fine-tune the IDS rules for appropriate thresholds so those teams don't get slammed with thousands of false positive alerts. IDSs are passive systems and only observe the network traffic, hence they do not have any effect on the network throughput.

- **Intrusion Prevention System** (**IPS**): An IPS detects and blocks malicious traffic. An IPS is required to be implemented in the line of traffic so it can prevent traffic from entering the network. This often reduces the throughput of the network.

Both an IDS and IPS can be set up either on the network layer, that is, between the outside network and the organization's network, or at the host level, that is, individually on user machines in the form of software applications.

In the next section, we will review the **Domain Name System** (**DNS**), which helps us visit websites without remembering the actual IP addresses.

The Domain Name System

The DNS is a critical component of the internet that acts as a directory for internet domain names and their corresponding IP addresses. Essentially, it acts as a telephone book for the internet, allowing you to associate a memorable and recognizable domain name (for example, `www.grcmusings.com`) with a series of numbers that make up the IP address (for example, `162.241.252.221`).

When you type a URL or domain name into your web browser, your computer sends a request to a DNS server to convert the domain name into an IP address. The DNS server then returns the IP address for the domain name, which the browser uses to connect to the website's server and display the website's content.

For example, if you type `www.grcmusings.com` into your web browser, the DNS server will translate the domain name into `162.241.252.221`, which is the GRCMusings site's IP address. The browser then uses this IP address to connect to the website's server, retrieve the content, and render it in the browser.

The DNS is a crucial component of the internet that allows users to access websites using memorable and recognizable domain names instead of having to remember a series of numbers (the IP address). Without it, users would have to type in IP addresses every time they wanted to access a website, making the internet much less user-friendly.

Now, you may try entering the IP address `162.241.252.221` to access `www.grcmusings.com` and it will throw an error. The reason for that is, often, multiple websites are hosted on the same server and it is impossible for the server to return a website unless it knows the *host headers*, or the IP address may be to a load balancer and not the actual server. In any case, I highly encourage you to try accessing the website using the IP address and see the results for your own learning. Also, you can find the IP address of any website by using the `ping` command in Command Prompt, as shown in the following figure:

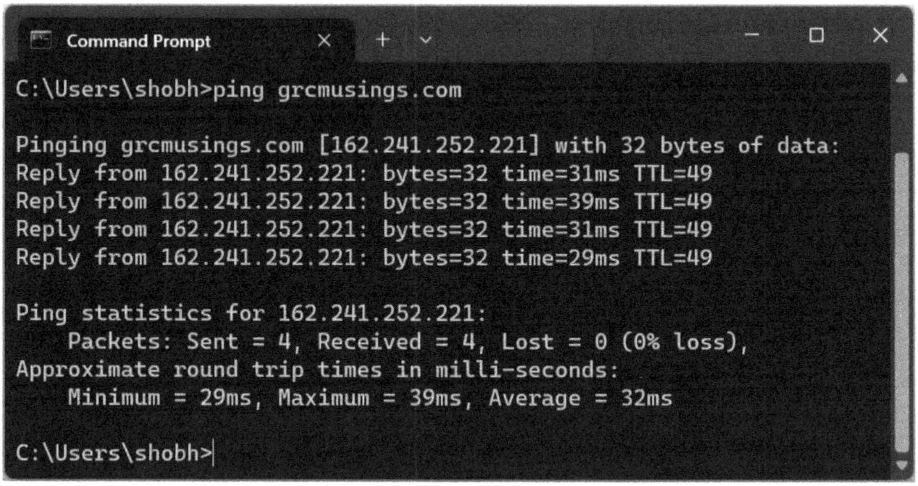

Figure 14.3 – Pinging GRCMusings.com

The DNS is an application layer protocol that operates at Layer 7 of the OSI model.

We will discuss the workings of a wireless network in the following section.

Wireless networks

Wireless networks or Wi-Fi networks are computer networks that use wireless communication to transmit data between devices. Unlike traditional wired networks that use cables to connect devices, wireless networks use radio waves to transmit data over the air. This allows devices such as laptops, smartphones, tablets, and other Wi-Fi-enabled devices to connect to the internet and exchange data without the need for physical cables.

Wireless networks are typically created using a **wireless router**, which acts as a central hub for transmitting and receiving data. Wireless networks are also subject to certain security risks, such as hacking and unauthorized access or installation of a rogue access point that creates a wireless network (**wireless access point**, or **WAP**). To protect against these risks, it is important for the risk manager to secure the wireless network with strong encryption, such as **WPA2**, and frequently train employees on using secure networks while accessing corporate applications from public networks such as a library or coffee shop.

Virtual private networks

Imagine that you have a corporate device that you want to connect to a private network, such as your workplace network. Without a **Virtual Private Network** (**VPN**), you would need to physically connect to that network, either by being on-site or using a remote access tool that connects you to the network. This can be performed remotely with the help of a VPN. A VPN allows us to create a secure, encrypted tunnel over a less secure network, such as the public internet. The purpose of a VPN is to provide a way to connect to a private network, such as a workplace network, from a remote location. This is particularly useful for employees who work from home or while traveling and need to access sensitive company data without compromising the security of the data.

When you connect to a VPN, the device first establishes an encrypted connection to the **VPN server** that can be located anywhere in the world. Once this connection is established, the device then uses the VPN server as a gateway to access the internet, making it appear as though your device is located at the same location as the VPN server. This means that you can access websites and services that may be restricted in your location, such as streaming services or social media platforms that are banned in certain countries.

A VPN can be implemented by using the IPSec protocol on the network layer or the TLS protocol on the transport layer.

Cloud computing

Cloud computing enables users to access computing resources managed by a third party over the internet. It provides a scalable and flexible infrastructure for deploying and managing a wide range of applications and services without the need for on-premises hardware and infrastructure.

At the core of cloud computing is the concept of **virtualization**. Virtualization allows multiple users to share the same physical resources, such as servers, compute, network, and storage, by creating virtual machines or containers. This enables cloud providers to deliver computing resources on demand, and users to access these resources from anywhere with an internet connection.

Now, let's briefly look at the various cloud computing service models in the upcoming sections.

Cloud computing service models

Cloud computing primarily provides three service models that organizations can avail per their requirements:

- **Infrastructure as a Service** (**IaaS**): IaaS provides users with access to virtualized computing resources, such as servers and storage, that can be scaled up or down on demand. A couple of examples of IaaS providers are **Amazon Web Services** (**AWS**) and Microsoft Azure.

- **Platform as a Service (PaaS)**: PaaS provides a platform for building, deploying, and managing applications so that organizations do not have to maintain the underlying infrastructure. A couple of examples of PaaS providers are **AWS Elastic Beanstalk** and **Google App Engine**.

- **Software as a Service (SaaS)**: SaaS provides users with access to software applications without the need for on-premises installation and maintenance, leaving users to manage only the configuration of the application. A couple of examples of SaaS services are **Gmail** and **Office 365**.

Cloud computing deployment models

Cloud computing services offer four deployment models that are primarily applicable to the IaaS service model:

- **Public cloud**: Public cloud services are provided by third-party cloud providers and underlying resources are shared with other organizations. AWS and Microsoft Azure are examples that provide a comprehensive suite of cloud services, such as compute, data storage, networking, and AI capabilities, that is available to the public for building and deploying applications.

- **Private cloud**: Private cloud services are provided by third-party cloud providers, but the underlying resources are not shared with other organizations. VMware vSphere is such an example that allows organizations to create and manage their private cloud infrastructure within their own data centers, providing dedicated resources and control over security and compliance.

- **Hybrid cloud**: Hybrid cloud services combine public and private cloud services to provide a unified and integrated computing environment. An example of this would be Google Anthos, which allows organizations to run applications on-premises and in multiple public cloud environments, providing a consistent management and deployment experience.

- **Community cloud**: Community cloud services are a derivation of the private cloud that is only available to a specific business community, such as universities or hospitals. **Government Community Cloud (GCC)** by Microsoft is an example of a community cloud environment for government agencies that enables them to share resources, collaborate, and meet specific compliance requirements.

We learned about the cloud computing service models and deployment models in this section. In the following section, we will review the security considerations for using cloud computing technology.

Security considerations of cloud computing

Cloud computing has made a paradigm shift in the technology industry; however, it does not come without additional costs. The following are some security considerations that the risk manager should be aware of while using cloud computing:

- **Shared responsibility**: In most cloud computing arrangements, there is a shared responsibility model between the cloud provider and the customer for security. It's important to understand the specific responsibilities of each party and ensure that appropriate security measures are in place. The shared responsibility model from AWS is an excellent resource to explore this further: `https://aws.amazon.com/compliance/shared-responsibility-model/`.

- **Service-Level Agreements (SLAs)**: SLAs define the terms of service and support provided by the cloud provider. It's important to ensure that security is included in the SLA and that appropriate measures are in place to protect against data loss, downtime, and other security risks.

- **Data breaches**: The risk of data breaches and unauthorized access to sensitive information is a major concern in cloud computing. It's important to ensure that proper encryption, access controls, and monitoring are in place to protect against data breaches.

- **Compliance**: Many organizations must comply with various industry and government regulations regarding data security and privacy. It's important to ensure that the cloud provider is compliant with relevant regulations and that the appropriate security controls are in place to meet these requirements.

- **Disaster recovery**: A disaster recovery plan must be in place to protect against data loss and downtime in the event of a disaster. This includes regular backups, testing of the recovery process, and ensuring that appropriate measures are in place to protect against disasters.

- **Identity and access management**: To prevent unauthorized access to cloud resources, it's important to have strong identity and access management controls in place. This includes strong authentication, authorization, and enabling multi-factor authentication.

- **Vendor lock-in**: It's important to consider the potential for vendor lock-in when selecting a cloud provider and to ensure that appropriate measures are in place to mitigate the risk of vendor lock-in.

In the following section, we will summarize the learnings and review the relevance of these concepts to a risk manager.

Summary

At the beginning of this chapter, we learned about EA, CMM, and the importance of each for maturing the organization's EA and measuring the maturity of implemented processes and systems. After that, we looked at the components of technology architecture, such as networks, networking devices, wireless networks, firewalls, IPSs/IDSs, DNS, wireless networks, and VPNs, and their relevance as a backbone for building an IT operations center. Finally, we learned about cloud computing, cloud computing deployment models, and security considerations for implementing cloud computing.

In the next chapter, we will learn about enterprise resiliency and data life cycle management.

Review questions

1. Networking devices, storage, and software are components of:

    A. Business architecture

    B. Technology architecture

    C. Data architecture

    D. Application architecture

2. Which of the following is *not* a property of the *Defined* process of the Capability Maturity Model?

    A. Well characterized and understood

    B. Defined at the organizational level

    C. Proactive

    D. Unpredictable

3. Which of the following is Layer 3 of the TCP/IP model?

    A. Physical

    B. Data link

    C. Network

    D. Application

4. Which of the following is Layer 5 of the OSI model?

    A. Physical

    B. Application

    C. Session

    D. Transport

5. Which of the following technologies provides a secure tunnel to log in remotely to a corporate network?

    A. Intrusion Detection System

    B. Intrusion Prevention System

    C. Virtual Private Network

    D. Domain Network System

6. Which of the following protocols is used to implement a VPN?

   A. DNS

   B. IPSec

   C. SSL

   D. SSO

7. Which of the following is an example of on-demand compute, network, and storage computing services?

   A. **Platform as a Service (PaaS)**

   B. **Software as a Service (SaaS)**

   C. **Infrastructure as a Service (IaaS)**

   D. **Network as a Service (NaaS)**

8. An online platform for video games is an example of _____.

   A. PaaS

   B. SaaS

   C. IaaS

   D. NaaS

9. In which deployment model are underlying cloud services *not* shared with other customers?

   A. Public

   B. Private

   C. Community

   D. Hybrid

10. In which deployment model are underlying cloud services shared with similar customers?

   A. Public

   B. Private

   C. Community

   D. Hybrid

Answers

The following are the answers to the questions in the previous section:

1. **B**. The components mentioned in the question are an example of technology architecture.

2. **D**. *Unpredictable* is not a property of the *Defined* process.

3. **C**. The network layer is Layer 3 of the TCP/IP model.

4. **C**. The session layer is Layer 5 of the OSI model.

5. **C**. A VPN creates a tunnel between unprotected networks and the corporate network to enable secure access.

6. **B**. The IPSec protocol enables VPN implementation.

7. **C**. On-demand compute, network, and storage computing services are an example of IaaS.

8. **B**. An online video game platform that lets users play games without installing any software is an example of SaaS.

9. **B**. Computing resources are not shared with other customers in a private deployment model.

10. **C**. A community cloud deployment model lets similar organizations share the underlying cloud resources.

15

Enterprise Resiliency and Data Life Cycle Management

This chapter discusses how we can build the foundations of a resilient architecture and ensure effective data life cycle management. It is divided into two main parts – one on **enterprise resiliency** and one on **data life cycle management**. In the first part, we will review the concepts related to enterprise resiliency, business continuity, disaster recovery, and recovery objectives. These are important considerations for an organization, being related to infrastructure availability, scalability, and reliability as well as the external challenge of attacks from malicious actors. In the second part, we will review the concepts of data classification, labeling, and data life cycle management, as well as regulatory requirements for retention and destruction.

In this chapter, we will cover the following key topics:

- Enterprise resiliency
- Business continuity and disaster recovery
- Recovery objectives
- Data classification and labeling
- Data life cycle management

With that, let us dive into the first section on enterprise resiliency.

Enterprise resiliency

Organizations encounter both internal and external threats all the time. **Resiliency** refers to the ability of an organization to withstand these threats and disruptions with minimum harmful impact and recover to normal business operation quickly.

It is important to differentiate between **resiliency** and **recovery**, which are often treated the same. While resiliency deals with avoiding or mitigating failure in the first place and continuing to provide services, recovery deals with restoring services after a failure has already occurred. Similarly, **reliability**, that is, the ability of a service to operate at an expected level, is an outcome of a resilient system.

> Quick definitions
>
> *Resiliency*: Keeping the system from complete failure
>
> *Recovery*: Recovering data or applications after a failure has occurred
>
> *Reliability*: An outcome of a resilient system performing at expected levels

A risk manager should analyze the organization's requirements and work with the engineering teams to define and implement adequate controls for resiliency. There are two major characteristics of a resilient plan – **business continuity** and **disaster recovery**. Let's review these concepts in the next section.

Business continuity and disaster recovery

Business continuity (**BC**) refers to the ability of an organization to continue operating its critical business functions during and after a disruption. The purpose of BC planning is to identify potential threats to an organization's operations and develop strategies and procedures to ensure that critical business processes can continue or resume quickly after a disruption. To prepare against such a disruption or disaster, you use a formal plan known as a **business continuity plan** (**BCP**).

The inputs to creating a BCP result from a **business impact analysis** (**BIA**), that is, the determining systems, data, processes, people, and other assets that are strategically important to achieving business goals. The BIA conducted by the organization will support the risk practitioner in recommending a reasonable and appropriate risk response and guide senior management in selecting appropriate mitigation strategies.

Disaster recovery (**DR**) refers to the ability to restore the data and applications that run the business, that is, infrastructure such as the data center and servers, and how quickly data and applications can be recovered and restored. BC planning refers to the strategy that helps a business operate with minimal or no downtime or service outage. DR is the process of restoring an organization's IT infrastructure and operations following a disruptive event, such as a natural disaster or cyber attack. BC involves planning for and maintaining the organization's essential functions during a disruption, including staffing, facilities, the supply chain, and communication with relevant stakeholders and customers. Both DR and BC are important components of a comprehensive plan to ensure the organization can continue to operate even in the face of disruption. It can be said that DR is a subset of BC planning.

The following figure shows the disaster planning flow that will be followed in the case of an unforeseen event:

Figure 15.1 – Disaster planning flow

In the following section, we will look at the relationship between resiliency and developing a BCP.

Relationship between resiliency and the BCP

Resiliency is closely related to the BCP as both deal with the organization's ability to maintain critical functions during unexpected events. In essence, maintaining resiliency is the overarching goal of the BCP. If an organization succeeds in performing a thorough assessment of critical functions, develops a detailed plan outlining the steps to be followed during a disruption, and tests the plans on a regular basis to assess their effectiveness, it will automatically incorporate resiliency into its business operations.

> **Important note**
> Resiliency is the end result that enables the organization to respond to unexpected events and the BCP is the framework that enables organizations to build resiliency.

In the next section, we will review the importance of **recovery objectives**, which are the outputs of the BIA exercise and important components of DR planning.

Recovery objectives

As we saw in *Figure 15.1*, BC planning typically begins with a BIA. The goal of a BIA is to identify the critical systems and services and ensure that sufficient controls are put in place so they operate within their **recovery point objective** (**RPO**) and **recovery time objective** (**RTO**), which are two key metrics to determine the criticality of an application.

The RPO is the maximum amount of data that an organization can afford to lose without a material impact, whereas the RTO is the maximum amount of time an application can remain unavailable before having a material impact on the business.

A similar metric to the RPO is the **maximum tolerable downtime** (**MTD**), that is, the maximum amount of time stakeholders are willing to accept for a business process outage that includes all impact considerations.

These metrics are illustrated in the following figure:

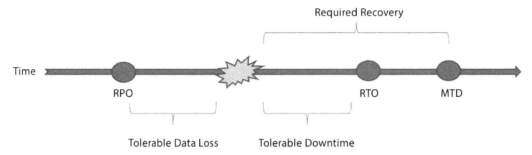

Figure 15.2 – The relationship between the RPO, RTO, and MTD

A risk practitioner should work closely with other teams to determine the RPO and RTO but is not required to determine these metrics for them. This is because business owners are the best assessors of what is most critical to them. The cost of maintaining a low RPO and RTO is inversely proportional to the recovery objectives; that is, the lower the recovery objectives, the higher the cost. Systems with a lower RPO will require frequent backups and recovery tests. Also, organizations that have systems with a low RTO will either need to deploy multiple servers to run the same services or establish parallel infrastructure to keep these metrics low.

Also, determining the RPO and RTO is only the initial part of BC planning; the risk manager should strive to conduct DR tests, such as restoring the database from a backup and serving a fraction of production traffic from the restored server, to ensure that backups are indeed being performed appropriately and can be used if the primary database gets corrupted or deleted.

These concepts are also discussed in detail in *Chapter 9* and I highly encourage you to review the examples provided. For now, we will move to the next part of this chapter, which deals with the first step of data management, that is, classifying data and labeling it.

Data classification and labeling

Data classification and labeling are integral parts of data life cycle management. The classification of data determines the sensitivity of the data and the controls that are required to keep it secure.

Data classification refers to the process of categorizing data based on the level of sensitivity and its value to an organization. Data classification determines the robustness of data controls to ensure security. The classification of data can be performed based on the following factors:

- **Regulatory requirements**: Data classification may be based on specific regulatory requirements. Regulatory and compliance frameworks such as HIPAA, GDPR, or PCI-DSS require certain types of data, such as **Personally Identifiable Information** (**PII**), to be classified and handled in a specific way to ensure compliance.

- **Business impact**: Data classification may be based on the level of impact that a loss or breach of the data could have on the business. Highly confidential financial data or intellectual property may be classified as *Critical*, whereas public information such as advertisements may be classified as *Low*.

- **Type**: Data classification may be based on the type of data, such as PII, financial data, health records, or intellectual property.

- **Access control**: Data classification may be based on the level of access required to view or modify the data. Data that requires special access privileges or privileged permission such as superuser rights may be classified as *Restricted*.

- **Location**: Data classification may be based on the physical location or storage of the data. For instance, any data that is stored on-premises within the country can be classified as *Critical* and data stored in the cloud in a different country can be classified as *Low*.

Data classification is only the first part of appropriate data management and is incomplete without the labeling or tagging of data. The data classification categories are defined by the risk manager and the data owner; that is, whoever creates the data is responsible for tagging or labeling the data. This involves actually putting a tag or a field in the document stating the classification of data. If the data classifications are not socialized within the organization and employees are not mandated to tag the data they own/create, the entire classification exercise will not be useful.

Another example of the importance of data labeling is in **data leakage prevention** (**DLP**). Almost all industry DLP tools either have a by-default classification based on the pattern of the data (such as nine digits for classifying social security numbers) or require users to label the data so the automated engine can flag it as appropriate.

In the next section, we will review data life cycle management and how classification/labeling plays an important role in it.

Data life cycle management

Data life cycle refers to the stages that data goes through, from its creation to its destruction. The following are the six stages of the data life cycle:

1. **Creation**: The creation stage refers to the phase when data first comes into existence or is collected through synthesis from other sources, such as consolidating a large amount of data into a report.

2. **Storage**: The storage stage refers to data being stored and thus becoming available for use.

3. **Use**: This stage refers to data being used or processed for a material output.

4. **Sharing**: This stage refers to data being shared with other users or entities for inference.

5. **Archiving**: When the data is no longer in use, it is archived to comply with regulatory/contractual requirements.

6. **Destruction**: Once the archival period is over or data is deemed inappropriate for usage, the data is destroyed using industry-standard best practices.

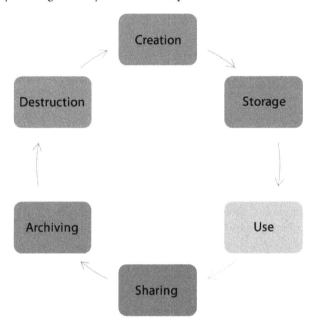

Figure 15.3 – The life cycle of data

Many of these stages are dependent on the classification and labeling of data. For instance, data that is classified as sensitive will have a policy for storage, use, sharing, and archiving that is separate from that of less sensitive data that is classified as public information. The more sensitive the classification of the data, the greater the controls and costs of securing it.

In the next section, we will compare data management and data governance.

Comparing data management and data governance

Data management and **data governance** are related concepts, but a risk manager should understand that they have different scopes and goals. The goal of data management is to facilitate the stages of the data life cycle and ensure that technical controls are implemented to ensure data is accurate, complete, available, and secure throughout the life cycle. On the other hand, the purpose of data governance is to ensure that data ownership, quality, security, and compliance are maintained.

> **Important note**
>
> Data management is focused on the *technical aspects* of data handling, whereas data governance is focused on the *strategic and policy aspects* of data management. Both are important for effective data management and organizations should have a clear understanding of both to manage their data effectively.

This brings us to the end of this chapter. In the next section, we will summarize and review what we learned in this chapter.

Summary

At the beginning of this chapter, we learned about the importance of enterprise resiliency and its relationship with BC and DR. In the following sections, we learned about the concepts of BIA and how it relates to the recovery objectives of the RPO, RTO, and MTD. In the final section, we learned about the difference between data classification and labeling, data life cycle management, and how data governance differentiates from data management. With the prevalence of cloud technologies, a risk manager should be aware of these concepts and plan with the business owners accordingly to reduce the risk of any outages or security issues that may impact the availability of data or systems.

In the next chapter, we will learn about the system development life cycle and emerging technologies.

Review questions

1. ___ refers to a system's ability to avoid complete failure.

    A. Recovery

    B. Resiliency

    C. Reliability

    D. None of the above

2. Which of the following acts as an input for the BCP?

    A. Recovery objectives

    B. DR test

    C. BIA

    D. BC test

3. A risk manager wants to develop a BCP and confirms that the application in scope can't lose more than four hours of data. Which recovery objective is the risk manager referring to?

    A. MTD

    B. RPO

    C. RTO

    D. BIA

4. An application can tolerate four hours of downtime but has to be recovered within six hours. Considering this, which of the following metrics is correct?

    A. RPO – four hours, RTO – six hours

    B. MTD – four hours, RTO – six hours

    C. MTD – six hours, RPO – four hours

    D. RTO – four hours, MTD – six hours

5. A risk manager defined the data classification for the entire organization. The organization processes healthcare records and often collects personal information such as driver's licenses to verify identity. What would the most appropriate classification for this data be?

    A. Sensitive (includes PII and PHI)

    B. Public

    C. PII

    D. PHI

6. Which stage of data life cycle management includes the disposal of data?

    A. Destruction

    B. Archival

    C. Use

    D. Storage

7. Data labeling is important for which of the following?

    A. Data archival

    B. Data destruction

    C. DLP

    D. All of the above

Answers

1. **B.** A system that can avoid complete failure is a resilient one.

2. **C.** Business impact analysis, or BIA, is the input to creating a BCP.

3. **B.** The correct answer is RPO, as the metric in question is data that can be lost.

4. **D.** The correct answer is an RTO of four hours and an MTD of six hours.

5. **A.** As the organization contains both PII and PHI, the correct classification is sensitive.

6. **A.** The destruction of data is also known as the disposal of data.

7. **D.** Data labeling is important to facilitate all the options.

16

The System Development Life Cycle and Emerging Technologies

The world of technology is constantly evolving and businesses must keep up with the latest developments to stay competitive. In this chapter, we will explore two critical topics that are essential to any successful organization – the **system development life cycle (SDLC)** and emerging technologies. The SDLC is a systematic approach to developing software applications that has been used for decades. It outlines the steps that must be taken to design, develop, test, and deploy a software system. Understanding the SDLC is crucial for any organization that wants to create high-quality software that meets its business needs. In the following section, we will discuss emerging technologies that are changing the way businesses operate. Technologies such as **blockchain**, **artificial intelligence (AI)**, **Internet of Things (IoT)**, and **quantum computing** have the potential to transform industries and create new opportunities.

The aim of this chapter is to understand the components of the SDLC and build a foundational understanding of emerging technologies. We will cover the following topics:

- Introducing the SDLC
- Project risk and SDLC risk
- System accreditation and certification
- Emerging technologies

With that, let us dive into the first section, in which we will learn about the SDLC.

Introducing the SDLC

The SDLC is a systematic process that outlines the steps involved in the development of software applications or information systems. The life cycle starts with **initiation**, followed by **development/acquisition**, **implementation**, **operation/maintenance**, and then **disposal**. Each phase has its own set of deliverable objectives and the process is designed to ensure that the software or system is delivered on time and within budget and meets the requirements of the stakeholders. The SDLC is essential in software development because it helps to manage the project effectively and minimize the risk of failure. By following a structured approach, the project team can ensure that the end product is of high quality and meets the needs of the stakeholders as well as users.

Phases of the SDLC

There are five phases of the SDLC. Let's briefly discuss each of them, along with the associated risks:

- **Phase 1 – initiation**: The initiation phase of the SDLC is the first phase of the process, where the project's feasibility is evaluated and the goals and objectives of the project are defined. During this phase, the project team identifies the stakeholders, gathers their requirements, and creates a project charter that outlines the scope, timeline, budget, and risks associated with the project. The initiation phase sets the foundation for the rest of the project and helps to ensure its success. Unclear project goals, lack of stakeholder support, inadequate budget or resources, and inaccurate project requirements are some of the risks associated with the initiation phase. The project team can increase the likelihood of delivering a successful project by addressing these risks during the initiation phase.

- **Phase 2 – development/acquisition**: The development/acquisition phase of the SDLC is where the actual coding of the software system takes place. The development team creates the software based on the specifications outlined in the first phase. If the software is being acquired, this phase involves selecting and implementing the software solution that best meets the organization's requirements. This phase determines whether the software will function as intended and meet the needs of the stakeholders. Poor-quality code, missed deadlines, and budget overruns are some of the risks associated with this phase. Additionally, the software may not meet the needs of the users or may be difficult to maintain in the long run. It is important to have a robust testing and quality assurance process in place during this phase to minimize the risks and ensure that the final product is of high quality and meets stakeholders' requirements.

- **Phase 3 – implementation**: The implementation phase includes installing the software system and training the users. This phase also includes configuring, enabling, testing, and verifying security configurations. The system not meeting the requirements of the users is a major risk of this phase as well. This can happen if the initiation phase was not thorough enough or if the developers did not fully understand the needs of the stakeholders. Another risk is that the implementation may take longer than expected, causing delays in the project schedule and budget overruns. To mitigate these risks, it is essential to have a well-defined implementation plan, clear communication with stakeholders, and a rigorous testing process. The implementation phase requires careful attention to detail to ensure that the system meets the needs of the users and is delivered on time and within budget.

- **Phase 4 – operation/maintenance**: In this phase, the software or system is deployed and enters into a period of operation and maintenance. The focus of this phase is to ensure that the system is functioning as intended and any issues that arise are addressed promptly. The main activities in this phase include monitoring, bug fixes, updates, and enhancements. The risks associated with this phase include the potential for security breaches, data loss, and system failures. It is essential to have a well-defined maintenance plan in place to mitigate these risks and ensure that the system remains stable and secure.

- **Phase 5 – disposal**: This is the final stage in the SDLC, where the software system is removed from operation and is no longer needed. This phase includes the secure destruction of data and hardware and software components that are no longer required. The main objective of this phase is to ensure that sensitive data is destroyed in a secure manner and that no unauthorized access to the system can be obtained. Associated risks during the disposal phase include the potential for data breaches or leaks leading to significant financial and reputational losses for an organization. There may be legal and regulatory requirements that need to be met, such as data privacy laws, which can carry significant penalties if not followed. Another risk is the possibility of a security breach if the system or software is not adequately decommissioned, resulting in unauthorized access to the system or software, leading to the exposure of sensitive information.

In the next section, we will delve deeper into the specifics of project risk and SDLC risk.

Project risk and SDLC risk

Project risk and **SDLC risk** are two distinct types of risk that organizations must manage during the software development process.

Project risk refers to the risk associated with achieving the project objectives, such as delivering the project on time, within budget, and meeting the requirements of the stakeholders. Project risks can include **external factors**, such as changes in market conditions, as well as **internal factors** such as delays in the development process or unforeseen technical issues. Managing project risks involves identifying potential risks, assessing their likelihood and impact, and developing a plan to mitigate them.

On the other hand, SDLC risk refers to the risk associated with the development process itself. SDLC risks can include issues with requirements gathering, software design, coding errors, and testing. Managing SDLC risks involves implementing best practices, such as adhering to industry standards and following a structured development methodology to minimize the risk of failure.

> **Important note**
> Project risk and SDLC risk are interrelated. While project risk focuses on achieving project objectives, SDLC risk focuses on the development process itself. For example, if there are issues with the SDLC, such as a lack of testing, this could lead to delays in project delivery or even project failure.

Managing both project risks and SDLC risks is critical to the success of software development projects. By identifying potential risks, assessing their impact, and developing a plan to mitigate them, organizations can ensure that their software development projects are completed on time, within budget, and meet the requirements of the stakeholders.

In the next section, we will briefly look at system accreditation and certification.

System accreditation and certification

System accreditation and certification are two related but distinct processes that are often used to ensure the quality and reliability of systems and products.

Accreditation is the process of evaluating and verifying that a system or organization meets certain standards and criteria. This can involve a review of policies, procedures, and practices to ensure they comply with industry best practices or regulatory requirements. Accreditation may be conducted by internal auditors or external third-party auditors who specialize in conducting audits and assessments.

Certification is the process of issuing a formal document or certificate that attests to the fact that a product, service, or system meets certain predefined standards. It is conducted by a third-party organization that has been authorized to issue certifications. In many cases, certification is required before a product can be sold or a service can be offered to the public.

Both accreditation and certification are important processes for ensuring the quality and reliability of systems and products. They can help increase customer confidence, improve product quality, and provide a competitive advantage; however, it's important to note that accreditation and certification are not the same and that each process has its own specific requirements and procedures.

> **Important note**
>
> Accreditation can be performed by an internal auditor or an external third-party auditor. Certification is *always* performed by an external third-party auditor.

In the next section, we will learn about a few emerging technologies and the risks associated with each.

Emerging technologies

Emerging technologies have the potential to revolutionize the way we live, work, and interact with each other. Some of the emerging technologies that are likely to have a significant impact on the world in the near future include AI, IoT, and blockchain. Risk managers should keep themselves abreast of these latest trends and technological innovations to support the business by mitigating the potential risks when the organization decides to implement such a technology. In the following sections, we discuss some of these technologies, the associated risks, and some controls that the risk manager should be aware of as these emerging technologies become mainstream.

Bring your own device (BYOD)

BYOD refers to the policy of allowing employees to use their personal devices to access company resources such as networks and data. BYOD has several benefits for both employees and employers.

For employees, it offers greater flexibility and convenience, allowing them to work on their personal machines. It also enables them to use devices that they are already familiar with, which can lead to increased productivity and job satisfaction. For employers, BYOD can lead to cost savings by reducing the need to purchase and maintain company-issued devices.

However, BYOD comes with its own challenges and risks, and risk managers must be mindful of the controls required for security and data protection before allowing such an organization-wide policy. For example, risk managers should build policies and technical controls to ensure that sensitive data is not compromised when accessed from personal devices. Risk managers should also ensure that employees are trained on the proper use and handling of company data on their personal devices.

Internet of Things

IoT devices are equipped with sensors and software that allow them to collect and transmit data over the internet, enabling real-time monitoring and control of physical objects and environments, all without the intervention of a human.

However, the widespread adoption of IoT technology also raises concerns about data privacy and security. IoT devices are also vulnerable to cyber-attacks and the data they collect can be sensitive and personal. There have already been numerous incidents of IoT devices such as Amazon Alexa and Google Nest infringing on people's privacy. It is important for IoT manufacturers and users to prioritize data privacy and security measures to ensure that these technologies are deployed in a safe and responsible manner.

Artificial intelligence

AI is the science of developing intelligent machines that can perform tasks that typically require human intervention. AI is already used in a variety of applications, such as autonomous vehicles, customer service chatbots, and fraud detection systems. At the time of writing, a company named OpenAI has released GPT-4, which has revolutionized the AI industry.

AI has the potential to revolutionize a variety of industries, including healthcare, finance, and transportation, by providing new levels of efficiency and accuracy. However, the most important aspects of making AI mainstream are the potential risks and challenges that must be addressed to ensure that AI is deployed in a safe and responsible manner. Some of these challenges include ensuring data privacy and security, preventing bias in AI algorithms, and ensuring that AI is transparent and understandable to humans.

Blockchain

Blockchain is a distributed ledger technology that allows secure and transparent transactions without the need for intermediaries. It was originally developed for use in cryptocurrencies such as Bitcoin, but its potential applications have expanded to many other industries. Blockchain has already had a significant impact on the finance industry and its uses in healthcare and supply chain management are being extensively researched.

Blockchain technology is still in the early stages of development and there are issues around scalability, energy consumption, and regulatory uncertainty that need to be addressed. Despite these challenges, many experts believe that blockchain has the potential to revolutionize the way we do business and interact with each other in the digital age.

Quantum computing

Quantum computing is a nascent discipline that leverages the principles of quantum mechanics to execute intricate computations much more rapidly than classical computing. It leverages the characteristics of quantum bits, or qubits, which can exist in a superposition of multiple states simultaneously and can be entangled with other qubits to perform intricate computations.

Cryptography stands out as a promising area for the application of quantum computing, as it has the potential to decrypt numerous encryption methods that are presently in use.

These emerging technologies have the potential to revolutionize a variety of industries. However, as with any new technology, there are also potential risks and challenges that must be addressed to balance the dichotomy of technical revolution and using these technologies in a safe and responsible manner.

Summary

In this chapter, we explored both the SDLC and emerging technologies to get a comprehensive understanding of how technology can be developed and utilized to meet the needs of modern businesses. We learned about the various phases of the SDLC, including initiation, development, implementation, maintenance, and disposal. As a risk manager, you will be responsible for knowing about these stages and closely working with the SDLC team to ensure that sufficient controls are implemented at each stage. We also learned about the emerging technologies that are changing the world, such as BYOD, IoT, AI, blockchain, and quantum computing. These technologies have the potential to significantly impact several industries, and risk managers will need to adapt to stay relevant in this rapidly evolving field.

In the next chapter, we will learn about information security and privacy principles.

Review questions

1. A project is budgeted in which phase of the SDLC?

    A. Development

    B. Disposal

    C. Initiation

    D. Maintenance

2. End user training is an integral part of which phase of the SDLC?

    A. Development

    B. Disposal

    C. Initiation

    D. Implementation

3. Which of the following is the final stage of the SDLC?

    A. Development

    B. Disposal

    C. Initiation

    D. Maintenance

4. The key difference between project risk and SDLC risk is that ____.

    A. Project risk relates to project objectives, while SDLC risk relates to time.

    B. Project risk relates to project objectives, while SDLC risk relates to development objectives.

    C. Project risk relates to development objectives, while SDLC risk relates to project objectives.

    D. There is no difference; both are the same.

5. An organization would like to achieve a certification of compliance for its product. Which of the following is a must to achieve certification?

    A. A risk manager

    B. An internal audit team

    C. Policies

    D. An external audit firm

6. The practice of allowing employees to use personal laptops to access organizational resources is called ____.

    A. Remote working

    B. BYOD

    C. Virtual computing

    D. Cloud computing

7. Which of the following has the potential to break current encryption algorithms?

    A. AI

    B. Blockchain

    C. Quantum computing

    D. IoT

Answers

1. **C**. The project is budgeted in the initiation phase of the SDLC.

2. **D**. The end users are trained in the implementation phase of the SDLC.

3. **B**. Disposal is the final phase of the SDLC.

4. **B**. Project risk relates to project objectives whereas SDLC risk relates to development objectives.

5. **D**. An external audit firm is required to certify a product.

6. **B**. BYOD allows employees to use personal laptops to access organizational resources.

7. **C**. The leaps in quantum computing could break, that is, decrypt, the current encryption algorithms.

17

Information Security and Privacy Principles

This is the final chapter of the CRISC syllabus, where we'll learn about **information security** and **privacy** concepts. Information security involves protecting information from unauthorized access, use, retention, disclosure, disruption, modification, or destruction, while privacy refers to an individual's right to control their personal information. Both are essential for maintaining trust with users.

The principles of information security include confidentiality, integrity, availability, accountability, and non-repudiation. We will learn about each of these, along with the additional concepts of encryption, hashing, digital signatures, and so on that enable achieving these principles. The principles of privacy include respecting user rights for confidentiality and giving them the option to exercise these rights.

The aim of this chapter is to understand information security and privacy principles that secure the system and build trust with the users.

In this chapter, we will cover the following topics:

- Fundamentals of information security
- Access management
- Encryption
- Hashing
- Digital signatures
- Certificates
- **Public key infrastructure (PKI)**
- Security awareness training
- Principles of data privacy
- Comparing data security and data privacy

With that, let us dive into the first section on the fundamentals of information security.

Fundamentals of information security

Confidentiality, **integrity**, and **availability** (collectively known as **CIA**) are the fundamental pillars of information security. It is an absolute requirement for the risk manager to understand and account for these to ensure all decisions are risk-based and derived with these three security pillars in mind. Let's look at these pillars in detail:

- **Confidentiality**: Confidentiality ensures that information is only accessible to authorized individuals. Unauthorized access to sensitive information can lead to incidents such as identity theft, fraud, or damage to an individual's or organization's reputation. Confidentiality can be ensured through technical and administrative controls such as encryption, masking, access control, training, and other similar controls.

  Two important principles are related to maintaining confidentiality, as follows:

  - The **need-to-know principle** ensures that individuals should be given access only to information that is needed for them to perform their job functions. For example, when you enter a password on an e-commerce website, that password is hashed (more on hashing in the following section), and ideally, no one on the e-commerce company's team should have access to your password as that is not required for them to perform their job.

  - The **least privilege principle** dictates that individuals should have the minimum access required to perform their job functions. For example, a customer support agent on that e-commerce website may be able to see the orders you placed but will not be able to update or add an order on your behalf.

  Largely, the need-to-know principle refers to the limitations on the amount of data that should be available to relevant individuals, whereas the least privilege principle refers to the minimum amount of access in an application that should be available to perform certain jobs.

- **Integrity**: Integrity refers to maintaining the accuracy and completeness of information and ensuring that it has not been altered or tampered with. Data integrity is essential to ensure that information is reliable and trustworthy. Organizations can ensure data integrity by implementing technical controls such as hashing, access controls, and data validation. The integrity of data can be achieved through hashing, as we will see in the following sections.

- **Availability**: Availability ensures that information is accessible to authorized users when needed. This is important for ensuring that business operations can continue smoothly without interruption. Availability can be ensured through measures such as redundancy, backups, and sharding.

These pillars are illustrated in the following diagram:

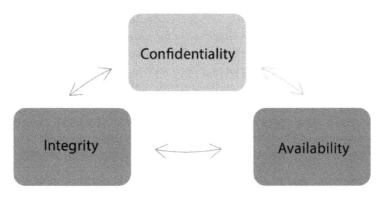

Figure 17.1 – The CIA triad

Non-repudiation is not a part of the CIA triad but often goes hand-in-hand with it. Non-repudiation is the ability to prove the origin and authenticity of a message to trace responsibility and enforce accountability for the sent message. Non-repudiation ensures the sender cannot deny sending a message later once it has been sent.

An example of non-repudiation is the use of digital signatures in email communication. When a sender digitally signs an email, they use a cryptographic algorithm to generate a unique digital signature based on the contents of the message. This signature can be verified by the recipient using the sender's public key, which confirms that the message was indeed sent by the sender and has not been altered in transit. We will learn about this in detail in the *Digital signatures* section later in this chapter.

In the next section, we will learn about the importance of access management and the different pillars for mature access management.

Access management

Managing access to information systems is one of the most integral parts of information security. The following four principles are the pillars of robust access management and are also known as **IAAA**:

1. **Identification**: Identification is the process of identifying an individual attempting to access a resource, information, or application. This is done by providing a unique ID such as a username, an email address, a user ID, or a similar attribute that can be easily remembered by the user. A risk practitioner should verify that the process of issuing this unique ID is secure and not shared with multiple users unless warranted for legitimate use such as a system/service account.

2. **Authentication**: After the identity of the user is verified, authentication is the process of verifying that they are who they claim to be. This is typically done by providing using one of the three factors—something you know (such as PIN, password), something you have (a security token, proximity card), or something you are (biometric authentication such as retinal scan or fingerprint).

The combination of an **identification** (username) and **authentication** (password) is collectively known as a **credential**. Only using a single factor for authentication such as a password poses an inherent risk because more often than not, the identification factor is either internal (such as your employee ID) or known to a lot of people (such as your email ID). Therefore, the onus of securing the account lies entirely on the strength of the second factor of the credential—that is, the authentication mechanism such as a password. By incorporating a second factor of authentication such as password and biometric, or password and a token, the organization can establish strong authentication that is difficult to break. This arrangement of authentication where multiple factors are used is known as **multi-factor authentication (MFA)**.

3. **Authorization**: After an individual has been authenticated, authorization is the process of determining which resources or information they are allowed to access. This is determined by the user's role or privileges within the organization. For example, assuming you are part of the information security team in your organization, you will have access to view your own pay stubs but not the pay stubs of your colleagues as you are not *authorized* to access that information. Principles of **least privilege** and **role-based access control** (**RBAC**) play an important role in implementing this principle in practice.

4. **Accountability**: Accountability is the process of logging and auditing who accessed which resources and when. This is done through log files or audit trails that can help with **incident response** (**IR**), compliance, and other legal requirements. A risk manager should ensure that audit logs are immutable so that even administrators of the system cannot change the logs, and even read-only access to these logs should be strictly controlled as the logs may have some confidential information.

An integral part of proper access management is **Segregation of Duties** (**SOD**). SOD is a principle in information security that aims to prevent internal fraud and other types of malicious activities triggered primarily by malicious insiders. SOD is implemented in organizations to ensure that no single individual has complete control to access critical data or applications.

The SOD principle is based on the idea that critical functions should be divided among different individuals to ensure that no one person can completely harm the system or commit fraud. For example, in a finance department, the person responsible for processing invoices should not also be responsible for approving them for payment, as this could lead to a conflict of interest and potentially allow for fraud or errors. This is also an important requirement for organizations to meet compliance requirements for various regulatory frameworks such as the **Sarbanes-Oxley Act** (**SOX**) or the **Payment Card Industry Data Security Standard** (**PCI DSS**). In cases where an organization doesn't have the resources to hold the SOD principle, the risk should be assessed periodically and accepted by the management in the form of an exception. Monitoring of activity logs, periodic user access reviews, and external audits are considered adequate compensating controls in these scenarios.

The next few sections will cover the basics of cryptography—that is, encryption, hashing, digital signatures, certificates, and PKI.

Encryption

Encryption is a process of converting plain text into cipher text with the help of mathematical algorithms and keys. The primary purpose of encryption is to protect sensitive data and ensure its confidentiality; however, it also supports achieving other aspects of security, such as integrity and availability.

Encryption works by taking a plain text message and using an encryption algorithm to convert it into cipher text. This cipher text can only be deciphered using a key that is known only to the intended recipient(s). The key is nothing but a very long prime number that is generated by the encryption algorithm or by a key management system.

Types of encryption

There are two main types of encryption, as follows:

- **Symmetric encryption** uses a single key for both encryption and decryption. It is faster and more efficient than asymmetric encryption, but it requires the key to be securely shared between the sender and the recipient. The following diagram illustrates the process:

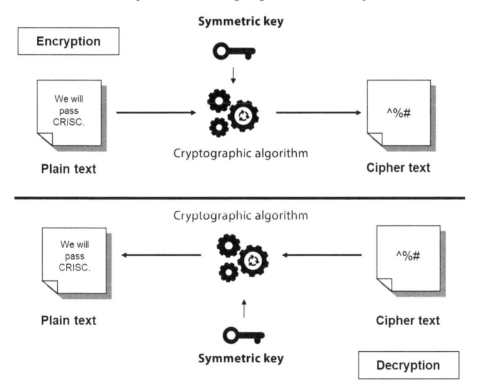

Figure 17.2 – Symmetric key encryption

- **Asymmetric encryption** uses a pair of keys—a **public key** and a **private key**—for encryption and decryption. This encryption method is also called public key encryption. It is slower but more secure as it uses different keys for encryption and decryption. Because of this reason, asymmetric keys are used only to encrypt short messages. The most common use of asymmetric key encryption is to share the key for symmetric encryption.

The following diagram shows asymmetric encryption in action, using a public and a private key:

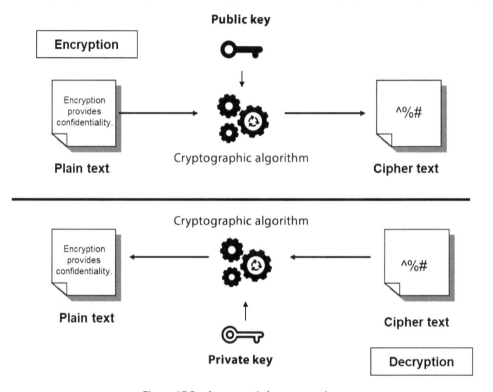

Figure 17.3 – Asymmetric key encryption

Encryption can also be used in combination with other security mechanisms, such as digital signatures and hashing, to provide additional layers of security. In the next section, we will review how encryption can be used with hashing.

Hashing

Hashing is the mathematical process to convert arbitrary-size input to a unique fixed-size output, also known as a hash or digest.

Hashing algorithms are one-way functions, meaning that it is computationally infeasible to reverse engineer the original input data from the hash output. Additionally, even minor changes to the input data will result in significant changes to the hash.

The practical applications of hashing range from password storage to data integrity checks and digital signatures. For example, in the case of password storage, the database stores a hash of the password instead of the actual password. When a user attempts to log in, their entered password is hashed and compared to the stored hash, and access is granted only after the two hashes match. There are additional requirements of adding *salt* and *pepper* to the input to prevent an attacker from guessing the password based on a pre-determined hash. Even though that is outside the scope of CRISC and this book, I highly encourage you to extend your learning of prevention from *rainbow table* attacks related to hashing. This is illustrated in the following diagram:

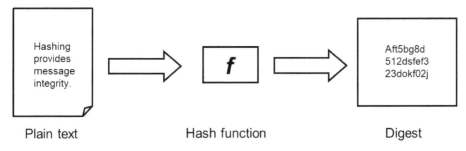

Figure 17.4 – Hashing

The following screenshot shows two SHA-256 hashes generated using an online tool at `https://emn178.github.io/online-tools/sha256.html`. As you can see, the hashes are completely different for the same sentence; can you spot what changed the hash?

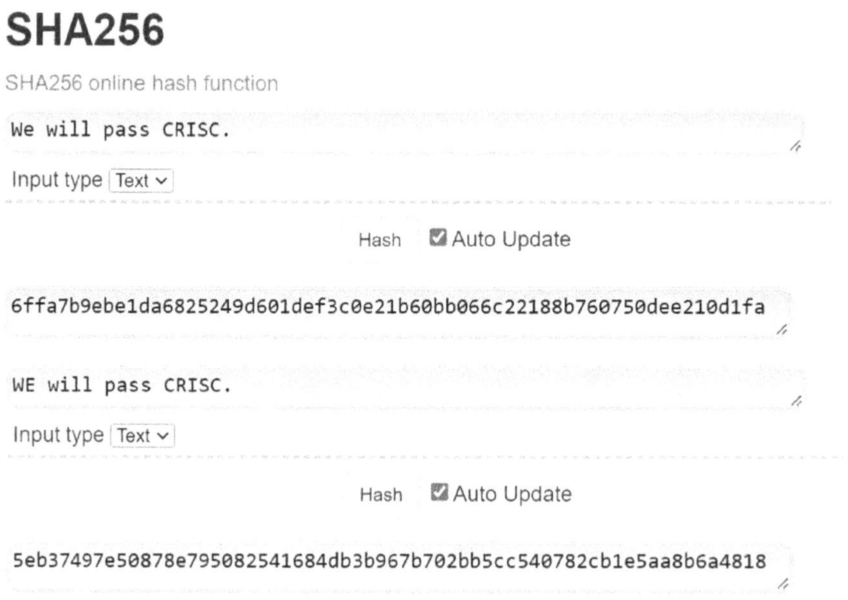

SHA256

SHA256 online hash function

We will pass CRISC.

Input type [Text ∨]

Hash   ☑ Auto Update

6ffa7b9ebe1da6825249d601def3c0e21b60bb066c22188b760750dee210d1fa

WE will pass CRISC.

Input type [Text ∨]

Hash   ☑ Auto Update

5eb37497e50878e795082541684db3b967b702bb5cc540782cb1e5aa8b6a4818

Figure 17.5 – SHA256 hashing in action

We will learn about the workings of digital signatures in the following section.

Digital signatures

Digital signatures provide the ability to verify the authenticity, integrity, and non-repudiation of electronic messages.

> **Important note**
> On their own, digital signatures do not provide confidentiality.

To create a digital signature, a sender uses a digital signature algorithm to generate a hash that is specific to the document or message. The sender then uses their private key to encrypt this value, creating the digital signature.

When the receiver receives the message, they use the sender's public key to decrypt the digital signature and obtain the hash value. They then hash the original message and compare the resulting hash value with the decrypted hash value. Considering the message has not been tampered with, the two hashes should ideally match.

In addition to providing authentication and integrity, digital signatures also provide non-repudiation, which means that the sender cannot deny having sent the message because the digital signature is unique to the sender's private key, which only they possess.

The following diagram illustrates a digital signature in action:

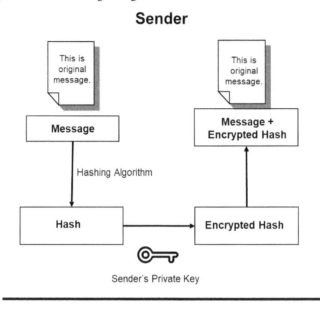

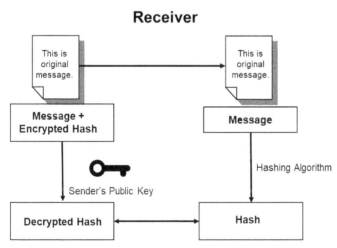

Figure 17.6 – Digital signature

It is important to note that digitally signing a message applies encryption only to the hash and not to the message itself, leaving the message unencrypted throughout the transmission. A digitally signed message could be any of the following:

1. **Encrypted and not signed**: Provides confidentiality, but not integrity or non-repudiation
2. **Signed and not encrypted**: Provides integrity and non-repudiation, but not confidentiality
3. **Signed and encrypted**: Provides confidentiality, integrity, and non-repudiation

Only messages that are signed and encrypted protect against unauthorized disclosure, maintain integrity, and ensure non-repudiation.

Certificates

The use of **public key encryption** (asymmetric encryption) allows for the decryption of data encrypted with the corresponding public key by the holder of a private key. However, it doesn't provide any confirmation of the identity of the person who owns the public key. To establish a link between public keys and specific owners, certificates are utilized with the aid of a trusted third party known as a **certificate authority** (**CA**). The CA confirms the identity of the owner using other methods and creates a certificate that the owner can use to verify that the public key belongs to them. By trusting this process, the receiver of a digitally signed message using the public key of the certificate can confirm that the message was sent and signed by the claimed sender. Additionally, the recipient can use the public key to encrypt a response that only the intended receiver can open.

These certificates can be issued by different CAs; imagine if each CA used its own format for the certificates—it would be impossible for browsers to understand all these formats. Therefore, certificates are based on the *X.509* standard that guarantees their accessibility across different systems, software, and web browsers, even if they are issued by different CAs. The certificate owner has the option to cancel the certificate by informing the CA, which then includes the certificate on a **certificate revocation list** (**CRL**). When a certificate validation request is made for a certificate on the CRL, the requester is alerted that the certificate has been revoked, indicating that it should not be considered trustworthy for identity verification.

Anytime we access `google.com` or `amazon.com`, we unknowingly use these certificates to verify that these websites are legitimate and that the identity of these websites is verified by the CA.

The following screenshot shows the *X.509* certificate for `amazon.com` issued by the **DigiCert Inc** CA. The certificate was initially issued on January 16, 2023, and is set to expire on January 16, 2024 (more often than not, these are renewed automatically):

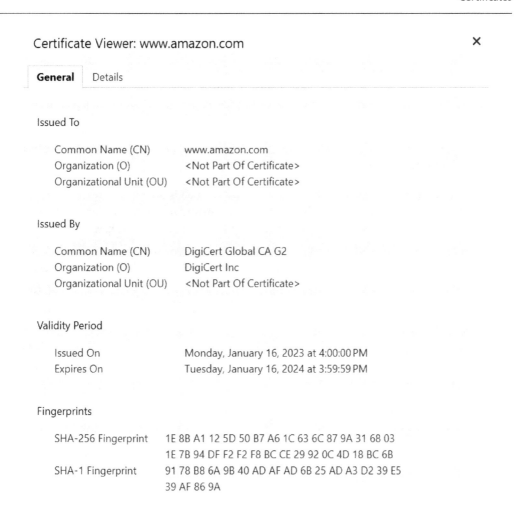

Figure 17.7 – X.509 certificate for amazon.com

Let's Encrypt is a CA that has excellent documentation on the working of certificates, which can be accessed here: `https://letsencrypt.org/how-it-works/`.

Public key infrastructure

A PKI is the overall implementation of certificates and CAs to establish, manage, distribute, and revoke digital certificates and public keys, which are used for authentication, encryption, and digital signatures. As the name suggests, a PKI relies on **public key cryptography**—that is, a pair of public and private keys.

There are many use cases for the implementation of a PKI; however, to understand it better, we'll take the example of https, as follows:

1. When you visit a website that has https in its URL, a PKI is used to secure the connection between your computer and the website. When you initiate a secure connection, your browser sends a request for the website's digital certificate that contains its public key.

2. The website responds with its digital certificate, and your browser uses the PKI to verify the certificate's authenticity, making sure it was issued by a trusted CA and that it has not been revoked.

3. Once the certificate is verified, your browser and the website use the public key to establish an encrypted connection that protects the data from unauthorized access or interception.

4. As we continue to interact with the website, the PKI is used to ensure that the communication remains secure and private. For example, if you submit your username and password on the website, the data you submit is encrypted using the website's public key and can only be decrypted by the website's private key.

The implementation of the digital signature that we saw in the *Digital signatures* section is also an example of a PKI.

The following screenshot illustrates the implementation of a certificate (an example of a PKI) on grcmusings.com:

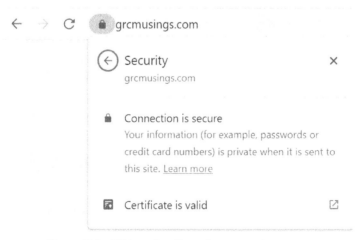

Figure 17.8 – PKI in action: "https" on grcmusings.com

In the PKI ecosystem, we know that the public key is distributed to everyone who needs to access a website, but who stores the private key? The private key is stored by the CA. If this private key is compromised, it will lead to a **single point of failure** (**SPOF**) for the entire PKI. Here is a complete timeline of such instances, leading to massive risk to all websites and systems using these CAs at that time: `https://sslmate.com/resources/certificate_authority_failures`.

In the following section, we will review the importance of security awareness training and how it can help create a culture of security across the organization.

Security awareness training

According to the *Data Breach Investigations Report* published by Verizon in 2022, "*82% of breaches involved the human element. Whether it is the use of stolen credentials, phishing, misuse, or simply an error, people continue to play a very large role in incidents and breaches alike.*" People, including insiders, pose a huge threat and the biggest risk of an incident or a breach. Therefore, it is imperative for organizations to prioritize training employees as much as protecting against external actors, and the best way to prevent internal threats is not by another tool or restriction on access but by continuously educating and upskilling employees in security-related matters.

Security awareness training is a critical component of any organization's cybersecurity strategy. It is designed to educate employees on how to identify and mitigate potential security risks and threats and to promote a culture of security throughout the organization. Attackers often target employees because when uneducated on security topics, they can be a weak link in the security chain, inadvertently introducing malware, falling for phishing scams, or sharing sensitive information. By providing regular security awareness training, ideally within 30 days of onboarding and at least annually thereafter, organizations can help employees become more vigilant and proactive in protecting themselves and the organization from these threats.

Security awareness training also helps to create a security-conscious culture within the organization. When employees understand the importance of security, they are more likely to adopt best practices and be proactive in reporting suspicious activities. This can help to create a strong security culture that reinforces the importance of security at all levels of the organization.

In the following section, we will review the principles of data privacy and how it impacts security and risk practitioners.

Principles of data privacy

Data privacy refers to the protection of the personal information of individuals or organizations from unauthorized access, use, or disclosure. Regulations around the world are cognizant of data privacy laws, and it is important for the risk manager to understand these principles that guide data privacy. Laws such as **General Data Protection Regulation** (**GDPR**) that apply to data stored anywhere in the **European Union** (**EU**) also set limits on transfers to other jurisdictions of data associated with EU citizens. The **California Privacy Rights Act** (**CPRA**) has a similar structure to GDPR but is applicable only to California residents, and many other state laws require companies to put equal emphasis on privacy as much as security. The following are some key principles that are relevant to data privacy:

- **Consent**: Organizations should obtain individual consent before collecting, using, or sharing personal information. Consent should be obtained through clear and specific language and recorded with the organization, and individuals should be provided the right to withdraw the consent at any time.

- **Accuracy**: Personal information should be accurate and up to date, and individuals should have the right to correct any inaccuracies.

- **Security**: Organizations should implement suitable security measures to protect their personal information against unauthorized access, disclosure, or use, including physical, technical, and administrative safeguards.

- **Accountability**: Organizations should be accountable for complying with data privacy laws and regulations and should provide individuals with access to their personal information and a means to request deletion. (*Note*—Apple now has a mandatory requirement for all applications that allow account creation within the app to add a button for account deletion as well: `https://developer.apple.com/support/offering-account-deletion-in-your-app`).

- **Transparency**: Organizations should be transparent about their data collection, use, and sharing practices. They should provide clear and easily accessible information in their privacy policies to individuals.

- **Confidentiality**: Organizations should maintain the confidentiality of personal information, especially sensitive information such as medical or financial information. The **Health Insurance Portability and Accountability Act** (**HIPAA**) in the US requires this from all covered entities and business associates for health-related information.

- **Privacy by design**: Organizations should integrate privacy considerations into the design and implementation of systems, products, and services. They should also conduct privacy impact assessments to identify and mitigate potential privacy risks.

- **Data breach notification**: Organizations should have policies, procedures, and template notifications in place to detect and respond to data breaches and notify affected individuals in a timely and transparent manner.

- **Data minimization**: Organizations should only collect the minimum amount of personal information necessary to achieve their stated purpose. They should avoid collecting additional information that is not needed or relevant to their operations.

- **Data retention**: Organizations should establish policies and procedures for the retention and deletion of personal information. They should only retain personal information for as long as it is needed for its stated purpose and should securely delete or anonymize it once it is no longer needed.

Organizations can enhance their data privacy practices and reduce the risk of data breaches or misuse of personal information by following these principles. Individuals can also use these principles as a guide to evaluate the privacy practices of organizations and to make informed decisions about sharing their personal information.

At this point, you may have a question—*What is the difference between data security and data privacy?* Let's look at this in the next section.

Comparing data security and data privacy

More often than not, risk practitioners consider data privacy and data security to be the same concepts; however, they are not. As discussed in the previous section, **data privacy** refers to the protection of personal information, ensuring that individuals have control over who can access their data, how it is used, and who it is shared with. On the other hand, **data security** refers to the measures taken to protect data from unauthorized access, use, disclosure, destruction, or modification.

Measures to ensure data privacy could involve obtaining explicit consent from individuals prior to collecting and using their personal data, implementing policies for retaining data, deleting that data when not required, and enabling individuals to access and manage their data. Measures to ensure data security could involve the use of access control, implementing strong password requirements, encrypting data in transit and at rest, implementing firewalls, and deleting data using approved deletion techniques and other technical controls to prevent unauthorized access, modification, or disclosure of data.

Data privacy emphasizes protecting the rights of individuals, while data security emphasizes protecting the confidentiality, integrity, and availability of data. In short, security enables privacy.

Summary

In this chapter, we learned about the fundamentals of information security—that is, **confidentiality** (make the information known only to intended parties), **availability** (keep the information available at all times for unauthorized users), and **integrity** (prevent unauthorized users from modifying the information), also known as the **CIA triad**. We then learned about the principles of access management, known in their abbreviated form as **IAAA**. These principles are **identification** (all users should be identifiable), **authentication** (all users should authenticate using **single-factor authentication** (**SFA**) or, preferably, MFA), **authorization** (users should only be able to perform operations that they are authorized to do), and **accountability** (all user activity should be logged and monitored).

In the next few sections, we learned about encryption and the types of encryption—symmetric (uses a single key for encryption and decryption) and asymmetric (uses a public/private key pair for encryption and decryption). We also learned about hashing algorithms that generate a one-way unique output for any amount of input, which can be used to generate digital signatures. Next, we learned about the use of certificates to legitimize the authenticity of a website and the importance of a PKI ecosystem in securing the internet. We then learned that malicious insiders pose the greatest threat to an organization and that the best control against it is the preventive control of providing them with security awareness training and developing a security-conscious culture within the organization. Finally, we discussed data privacy principles, why is it important for organizations to be cognizant of individual privacy, and the differences between data privacy and data security.

In the following chapters, we will bring all the learnings together and solve 200 practice questions. I strongly encourage you to read the descriptions of the correct answers, as well as the incorrect ones, to understand the rationale and ensure that you develop a concrete understanding of these topics for the exam.

Review questions

1. Which of the following keys are required for asymmetric key encryption?

    A. Only the public key

    B. Only the private key

    C. Any public or private key

    D. Public key and private key pair

2. Which of the following key(s) are required for symmetric key encryption?

    A. Public and private key

    B. A single key for encryption and decryption

    C. Only the public key and hash

    D. Only the private key and hash

3. Integrity of information means that ___.

    A. The information is available to authorized users only

    B. The information is available at all times to authorized users

    C. The information has not been tampered with

    D. All of these

4. Confidentiality of information means that ___.

    A. The information is available to authorized users only

    B. The information is available at all times to authorized users

    C. The information has not been tampered with

    D. All of these

5. Availability of information means that ___.

    A. The information is available to authorized users only

    B. The information is available at all times to authorized users

    C. The information has not been tampered with

    D. All of these

6. Which of the following access control principles speaks to the logging and monitoring of user activities?

    A. Identification

    B. Authentication

    C. Authorization

    D. Accountability

7. Which of the following access control principles requires each user to have a separate unique name?

    A. Identification

    B. Authentication

    C. Authorization

    D. Accountability

8. The SOD principle ensures that ___.

    A. All admin access is granted to a single user

    B. Access that can approve the transaction is distributed to multiple roles

    C. Confidentiality of information is maintained

    D. Identification of users is in place

9. Which of the following is true about hashing?

    A. Provides unique output for each input string

    B. Is non-reversible

    C. Helps maintain integrity

    D. All of these

10. Which of the following is *NOT* provided by digital signatures?

    A. Integrity

    B. Non-repudiation

    C. Authenticity

    D. Confidentiality

11. A sender would like to encrypt data such that everyone with access to the data can decrypt it using the sender's public key. Which of the following should the sender use to encrypt the data?

    A. Receiver's public key

    B. Sender's public key

    C. Receiver's private key

    D. Sender's private key

12. A sender would like only the receiver to decrypt data and no one else. Which of the following should the sender use to encrypt the data?

    A. Receiver's public key

    B. Sender's public key

    C. Receiver's private key

    D. Sender's private key

13. Which of the following cryptographic mechanisms should be used to provide confidentiality, integrity, and non-repudiation of a message?

    A. Signed, not encrypted

    B. Encrypted, not signed

    C. Signed, encrypted

    D. Signed, hashed

14. Which of the following could be a SPOF in the PKI ecosystem?

    A. Certificate

    B. CA

    C. Requester

    D. Browser

15. The *MOST* important aspect of conducting security awareness training is to ___.

    A. Train everyone on the latest threats

    B. Show compliance with external audits

    C. Create a security-conscious culture

    D. Upskill each employee to fill cybersecurity gaps

16. Data privacy should be ___.

    A. Each individual's right

    B. Supported by security controls

    C. Embedded into the system design

    D. All of these

Answers

1. **D.** Asymmetric encryption needs both a public key and a private key.

2. **B.** Symmetric encryption needs only one key for encryption as well as decryption.

3. **C.** Integrity means that information has not been tampered with at rest or in transit.

4. **A.** Confidentiality means that the information is available to authorized users only.

5. **B.** Availability means that the information is available at all times to authorized users.

6. **D.** The accountability principle deals with the logging and monitoring of user activities.

7. **A.** The identification principle requires all users to have a unique ID.

8. **B.** The SOD principle requires that no user should have sole access to approve all transactions.

9. **D.** All these options are true for hashing.

10. **D.** Digital signatures on their own do not provide confidentiality.

11. **D.** The sender should use their private key to encrypt the data so that the receiver(s) can use the sender's public key to decrypt the data.

12. **A.** The sender should use the receiver's public key to encrypt the data so that only the receiver can decrypt the data and read it. The receiver will use the sender's private key to decrypt the data.

13. **C.** A message has to be signed and encrypted to provide confidentiality, integrity, and non-repudiation.

14. **B.** The CA could be a SPOF if the private key of the CA gets compromised.

15. **C.** The most important goal of security awareness training is to create a security-conscious culture. All the other options are true as well but not the most important part of conducting such training.

16. **D.** Data privacy should be each individual's right, supported by security controls and embedded into the system design to enforce a **privacy-by-design** principle.

Part 6:
Practice Quizzes

In this part, you are provided with two practice quizzes that simulate the questions in the ISACA CRISC exam. The goal of this section is to determine your understanding of the concepts discussed throughout the book and also to review the explanations for both correct and incorrect answers. Before proceeding with this section, I recommend reviewing *Chapter 2, CRISC Practice Areas and the ISACA Mindset* again.

This part has the following chapters:

- *Chapter 18, Practice Quiz – Part 1*
- *Chapter 19, Practice Quiz – Part 2*

18
Practice Quiz – Part 1

1. Which of the following is not a component of the risk management process?

    A. Risk identification

    B. Risk analysis

    C. Risk acceptance

    D. Risk elimination

    Answer: **D**. Risk elimination.

    Risk elimination is *not* a component of the **risk management** process because it is not always possible to eliminate all risks, and attempting to do so may not be feasible. The other options – risk identification, risk analysis, and risk acceptance – are all important components of the risk management process.

2. What is the purpose of a risk assessment?

    A. To identify vulnerabilities in the system

    B. To determine the impact of a potential risk event

    C. To evaluate the effectiveness of existing controls

    D. All of the above

    Answer: **D**. All of the above.

    The purpose of a **risk assessment** is to identify vulnerabilities in the system, determine the impact of a potential risk event, and evaluate the effectiveness of existing controls. By performing a risk assessment, organizations can identify areas of weakness in their security posture and take steps to address them before they are exploited by attackers.

3. What is the purpose of a security policy?

    A. To establish the overall goals and objectives of the security program

    B. To define the roles and responsibilities of individuals within the organization

    C. To provide guidance on the acceptable use of information systems and data

    D. All of the above

    Answer: **D**. All of the above.

    The purpose of a **security policy** is to establish the overall goals and objectives of the security program, define the roles and responsibilities of individuals within the organization, and provide guidance on the acceptable use of information systems and data. Security policies are an essential component of any security program as they provide a framework for implementing and enforcing security controls.

4. Which of the following is not a type of access control?

    A. Administrative access control

    B. Physical access control

    C. Technical access control

    D. Social access control

    Answer: **D**. Social access control.

    Social access control is *not* a type of **access control**. The other options – administrative access control, physical access control, and technical access control – are all types of access control. **Administrative access control** refers to policies and procedures for managing user access, **physical access control** refers to physical barriers and controls to prevent unauthorized access, and **technical access control** refers to controls that are implemented in software and hardware to prevent unauthorized access.

5. Which of the following is not a category of information security controls?

    A. Technical controls

    B. Physical controls

    C. Administrative controls

    D. Legal controls

    Answer: **D**. Legal controls.

    Legal controls are *not* a category of **information security controls**. The other options – technical controls, physical controls, and administrative controls – are all categories of information security controls. **Technical controls** are implemented in software and hardware, **physical controls** are implemented in the physical environment, and **administrative controls** are policies and procedures for managing security.

6. What is the purpose of BIA?

    A. To identify critical business processes and functions

    B. To identify potential risks to the organization

    C. To determine the impact of a disruption on business operations

    D. All of the above

    Answer: **D**. All of the above.

    The purpose of BIA is to identify critical business processes and functions, identify potential risks to the organization, and determine the impact of a disruption on business operations. By performing BIA, organizations can prioritize their resources and focus their efforts on protecting the most critical assets.

7. Which of the following is not a type of risk response strategy?

    A. Risk avoidance

    B. Risk transfer

    C. Risk acceptance

    D. Risk elimination

    Answer: **D**. Risk elimination.

    Risk elimination is *not* a type of **risk response strategy**. The other options – risk avoidance, risk transfer, and risk acceptance – are all types of risk response strategies. **Risk avoidance** involves taking steps to avoid the risk altogether, **risk transfer** involves shifting the risk to another party, and **risk acceptance** involves accepting the risk and implementing controls to mitigate its impact.

8. What is the purpose of a vulnerability assessment?

    A. To identify vulnerabilities in the system

    B. To determine the likelihood of a potential risk event

    C. To evaluate the effectiveness of existing controls

    D. All of the above

    Answer: **A**. To identify vulnerabilities in the system.

    The purpose of a **vulnerability assessment** is to identify vulnerabilities in the system, such as software flaws, misconfigurations, or weak passwords. By identifying vulnerabilities, organizations can take steps to patch or mitigate them before they are exploited by attackers.

9. Which of the following is a component of a disaster recovery plan?

    A. Backup and recovery procedures

    B. Access control policies

    C. Security awareness training

    D. Performance monitoring tools

    Answer: **A**. Backup and recovery procedures.

    Backup and recovery procedures are critical components of a **disaster recovery plan**. In the event of a disaster, such as a natural disaster, a cyber-attack, or a system failure, organizations need to be able to recover their data and systems quickly to minimize downtime and prevent data loss. The other options – access control policies, security awareness training, and performance monitoring tools – are important components of a security program, but they are not specifically related to disaster recovery.

10. What is the purpose of a security incident response plan?

    A. To identify critical business processes and functions

    B. To identify potential risks to the organization

    C. To provide a framework for responding to security incidents

    D. To determine the impact of a disruption on business operations

    Answer: **C**. To provide a framework for responding to security incidents.

    The purpose of a **security incident response plan** is to provide a framework for responding to security incidents, such as cyber-attacks, malware infections, or system breaches. A well-designed incident response plan can help organizations detect and respond to security incidents quickly and effectively minimize the impact of the incident on business operations.

11. Which of the following is an example of a detective control?

    A. Firewall

    B. Intrusion detection system

    C. Encryption

    D. Access control list

    Answer: **B**. Intrusion detection system.

    An intrusion detection system is an example of a **detective control**. Detective controls are designed to detect potential security incidents or unauthorized access to systems or data. The other options – firewall, encryption, and access control list – are examples of **preventive or deterrent controls**, which are designed to prevent or discourage unauthorized access or security incidents.

12. Which of the following is a key concept of risk management?

    A. Risk assessment

    B. Data classification

    C. Security auditing

    D. Intrusion prevention

    Answer: **A**. Risk assessment.

    Risk assessment is a key concept of risk management. Risk assessment involves identifying and evaluating potential risks to the organization, and determining the likelihood and potential impact of those risks. Data classification, security auditing, and intrusion prevention are all important components of a security program, but they are not specifically related to risk management.

13. Which of the following is an example of a compensating control?

    A. Firewall rules

    B. Encryption

    C. Security awareness training

    D. Access control policy

    Answer: **B**. Encryption.

    Encryption is an example of a **compensating control**. Compensating controls are implemented to mitigate the risk of a vulnerability that cannot be fully remediated. For example, if a system has a vulnerability that cannot be patched, encryption can be used as a compensating control to protect data in transit or at rest. The other options – firewall, security awareness training, and access control policy – are not examples of compensating controls.

14. Which of the following is a benefit of implementing a security awareness program?

    A. Preventing malware infections

    B. Reducing the likelihood of data breaches

    C. Reducing the cost of security incidents

    D. Increasing network speed

    Answer: **B**. Reducing the likelihood of data breaches.

    Implementing a **security awareness** program can help reduce the likelihood of data breaches by educating employees about potential security risks and best practices for data protection. While security awareness training can help prevent malware infections and reduce the cost of security incidents, it is not directly related to increasing network speed.

15. Which of the following is a key component of a security governance program?

    A. Incident response planning

    B. Security awareness training

    C. Risk management

    D. User access management

    Answer: **C**. Risk management.

    Risk management is a key component of a **security governance** program. Security governance is the framework that defines the policies, procedures, and processes for managing information security risks within an organization. Incident response planning, security awareness training, and user access management are all important components of a security program, but they are not specifically related to security governance.

16. Which of the following is a benefit of implementing a security incident response plan?

    A. Reducing the cost of security incidents

    B. Increasing network speed

    C. Preventing malware infections

    D. Eliminating all security incidents

    Answer: **A**. Reducing the cost of security incidents.

    Implementing a **security incident response plan** can help reduce the cost of security incidents by enabling organizations to respond quickly and effectively to security incidents, minimizing their impact on business operations. While incident response planning can help prevent malware infections and improve network security, it cannot eliminate all security incidents.

17. Which of the following is an example of a technical control?

    A. Security policy

    B. Security awareness training

    C. Firewall

    D. Risk assessment

    Answer: **C**. Firewall.

    A firewall is an example of a **technical control**. Technical controls are designed to prevent, detect, or respond to security incidents using technology, such as firewalls, intrusion detection systems, and encryption. The other options – security policy, security awareness training, and risk assessment – are examples of **administrative or organizational controls**.

18. Which of the following is a key component of a risk assessment process?

    A. Developing security policies

    B. Identifying threats and vulnerabilities

    C. Implementing security controls

    D. Conducting security awareness training

    Answer: **B**. Identifying threats and vulnerabilities.

    Identifying threats and vulnerabilities is a key component of a **risk assessment** process. Risk assessment is the process of identifying, assessing, and prioritizing the risks to an organization's assets and determining the best way to manage those risks. Developing security policies, implementing security controls, and conducting security awareness training are all important components of a security program, but they are not specifically related to risk assessment.

19. Which of the following is an example of a physical control?

    A. Access controls

    B. Encryption

    C. Intrusion detection systems

    D. Security cameras

    Answer: **D**. Security cameras.

    Security cameras are an example of a **physical control**. Physical controls are designed to physically protect assets, such as buildings, equipment, and data centers, using measures such as locks, fences, and security cameras. Access controls, encryption, and intrusion detection systems are examples of **technical controls**.

20. Which of the following is a key component of a security risk management process?

    A. Developing security policies

    B. Conducting vulnerability scans

    C. Implementing access controls

    D. Conducting security awareness training

    Answer: **B**. Conducting vulnerability scans.

    Conducting vulnerability scans is a key component of a **security risk management** process. Risk management is the process of identifying, assessing, and prioritizing the risks to an organization's assets and determining the best way to manage those risks. Conducting vulnerability scans helps identify potential security vulnerabilities that could be exploited by attackers. Developing security policies, implementing access controls, and conducting security awareness training are all important components of a security program, but they are not specifically related to risk management.

21. Which of the following is a key component of an information security program?

    A. Defining security roles and responsibilities

    B. Developing software applications

    C. Creating marketing materials

    D. Managing human resources

    Answer: **A**. Defining security roles and responsibilities.

    Defining security roles and responsibilities is a key component of an **information security program**. An information security program is a set of policies, procedures, and practices designed to protect an organization's information assets from unauthorized access, use, disclosure, disruption, modification, or destruction. Developing software applications, creating marketing materials, and managing human resources are all important activities within an organization, but they are not specifically related to information security.

22. Which of the following is an example of a logical control?

    A. Security cameras

    B. Turnstiles

    C. Encryption

    D. Security guards

    Answer: **C**. Encryption.

    Encryption is an example of a **logical control**. Logical controls are designed to protect assets, such as data and software, using measures such as access controls, encryption, and intrusion detection systems. Security cameras and security guards are examples of **physical controls**.

23. Which of the following is a key objective of a security awareness training program?

    A. Eliminating all security incidents

    B. Increasing employee productivity

    C. Reducing the risk of security incidents

    D. Enhancing customer satisfaction

    Answer: **C**. Reducing the risk of security incidents.

    A key objective of a **security awareness training program** is to reduce the risk of security incidents by educating employees on how to recognize and respond to security threats. While a security awareness training program can improve employee productivity and enhance customer satisfaction, it cannot eliminate all security incidents.

24. Which of the following is a key component of a disaster recovery plan?

    A. Conducting vulnerability assessments

    B. Defining recovery time objectives

    C. Developing security policies

    D. Implementing access controls

    Answer: **B**. Defining recovery time objectives.

    Defining recovery time objectives is a key component of a **disaster recovery plan**. A disaster recovery plan is a set of policies, procedures, and practices designed to enable an organization to recover from a disaster or disruptive event. Conducting vulnerability assessments, developing security policies, and implementing access controls are all important components of a security program, but they are not specifically related to disaster recovery.

25. Which of the following is an example of a technical control?

    A. Password policies

    B. Security awareness training

    C. Access control lists

    D. Background checks

    Answer: **C**. Access control lists.

    Access control lists are an example of a **technical control**. Technical controls are designed to protect assets, such as data and software, using measures such as access controls, firewalls, and encryption. Password policies and security awareness training are examples of **administrative controls**, while background checks are an example of a **physical control**.

26. Which of the following is a key component of a security incident response plan?

    A. Conducting risk assessments

    B. Defining incident response procedures

    C. Developing security policies

    D. Implementing access controls

    Answer: **B**. Defining incident response procedures.

    Defining incident response procedures is a key component of a **security incident response plan**. A security incident response plan is a set of policies, procedures, and practices designed to enable an organization to respond to and recover from security incidents. Conducting risk assessments, developing security policies, and implementing access controls are all important components of a security program, but they are not specifically related to incident response.

27. Which of the following is the *best* example of a physical control?

    A. Turnstiles

    B. CCTV

    C. Picture IDs

    D. Alarm systems

    Answer: **A**. Turnstiles.

    All of these are examples of physical controls, but only turnstiles can perform access control for unauthorized users. CCTV and picture IDs are examples of physical controls but don't necessarily provide physical access control, while using alarm systems is very generic and not the best example of a physical access control.

28. Which of the following is a key objective of a security audit?

    A. Ensuring compliance with all laws and regulations

    B. Eliminating all security incidents

    C. Reducing the risk of security incidents

    D. Improving the efficiency of security controls

    Answer: **A**. Ensuring compliance with all laws and regulations.

    A key objective of a **security audit** is to ensure compliance with all laws and regulations applicable to an organization. While a security audit can help reduce the risk of security incidents and improve the efficiency of security controls, its primary purpose is to evaluate an organization's compliance with legal and regulatory requirements.

29. Which of the following is a key component of a security governance program?

    A. Conducting vulnerability assessments

    B. Developing security policies

    C. Defining security roles and responsibilities

    D. Implementing access controls

    Answer: **C**. Defining security roles and responsibilities.

    Defining security roles and responsibilities is a key component of a **security governance** program. A security governance program is a set of policies, procedures, and practices designed to ensure that an organization's information security practices are aligned with its business objectives and risks. Conducting vulnerability assessments, developing security policies, and implementing access controls are all important components of a security program, but they are not specifically related to governance.

30. Which of the following is a key responsibility of an information security manager?

    A. Developing security policies

    B. Conducting penetration test

    C. Implementing antivirus software

    D. Monitoring physical access

    Answer: **A**. Developing security policies.

    Developing security policies is a key responsibility of an **information security manager**. Information security managers are responsible for developing, implementing, and managing an organization's information security program, including its policies, procedures, and practices. Conducting penetration tests, implementing antivirus software, and monitoring physical access are not the key responsibilities of an information security manager.

31. Which of the following is an example of a technical vulnerability?

    A. Weak passwords

    B. Lack of security awareness training

    C. Physical theft of equipment

    D. Fire damage to a data center

    Answer: **A**. Weak passwords.

    Weak passwords are an example of a **technical vulnerability**. Technical vulnerabilities are weaknesses in an organization's technical infrastructure, such as software and hardware. Lack of security awareness training is an example of a human vulnerability, while physical theft of equipment and fire damage to a data center are examples of physical vulnerabilities.

32. Which of the following is an example of a security incident?

    A. A power outage

    B. A virus infection

    C. Accidental deletion of data

    D. A routine system upgrade

    Answer: **B**. A virus infection.

    A virus infection is an example of a **security incident**. A security incident is any event that poses a threat to an organization's information security. A power outage, accidental deletion of data, and routine system upgrades are not examples of security incidents.

33. Which of the following is an example of a logical control?

    A. Security cameras

    B. Access control lists

    C. Firewalls

    D. Security guards

    Answer: **C**. Firewalls.

    Firewalls are an example of a **logical control**. Logical controls are designed to protect assets, such as data and software, using measures such as firewalls, encryption, and access control lists. Security cameras and security guards are examples of **physical controls**, while access control lists are an example of a **technical control**.

34. Which of the following is the primary objective of a business continuity plan?

    A. Preventing incidents from occurring

    B. Responding to incidents when they occur

    C. Recovering from incidents when they occur

    D. Conducting risk assessments

    Answer: **C**. Recovering from incidents when they occur.

    The primary objective of a **business continuity plan** is to help an organization recover from incidents when they occur. A business continuity plan is a set of procedures and policies designed to ensure that essential business functions can continue in the event of a disruption. Preventing incidents from occurring is the objective of a risk management program, responding to incidents when they occur is the objective of an incident response plan, and conducting risk assessments is an important component of a risk management program.

35. Which of the following is the best approach to assess the effectiveness of security controls?

    A. Conducting periodic vulnerability scans

    B. Conducting a security audit

    C. Conducting a risk assessment

    D. Conducting a penetration test

    Answer: **B**. Conducting a security audit.

    Conducting a security audit is the best approach to assess the effectiveness of **security controls**. A security audit is a systematic evaluation of an organization's information security practices, including its policies, procedures, and technical controls. Conducting periodic vulnerability scans, conducting a risk assessment, and conducting a penetration test are all important components of an organization's information security program, but they are not specifically designed to assess the effectiveness of security controls.

36. Which of the following is the most important step in the risk management process?

A. Risk identification

B. Risk assessment

C. Risk treatment

D. Risk monitoring

Answer: **A**. Risk identification.

Risk identification is the most important step in the **risk management** process. Risk identification is the process of identifying potential risks to an organization's information assets. Without accurately identifying risks, it is impossible to effectively assess, treat, and monitor them. Risk assessment, risk treatment, and risk monitoring are also important steps in the risk management process, but they are built on the foundation of risk identification.

37. Which of the following is the primary goal of a security incident response plan?

A. To prevent security incidents from occurring

B. To detect security incidents as they occur

C. To respond to security incidents in a timely and effective manner

D. To recover from security incidents as quickly as possible

Answer: **C**. To respond to security incidents in a timely and effective manner.

The primary goal of a **security incident response plan** is to respond to security incidents in a timely and effective manner. A security incident response plan is a set of procedures and policies designed to help an organization respond to security incidents, including those that result in a loss or compromise of data or systems. Preventing security incidents is the objective of a risk management program, detecting security incidents is the objective of intrusion detection systems and other monitoring tools, and recovering from security incidents is the objective of a business continuity plan.

38. Which of the following is an example of a preventive control?

A. Intrusion detection systems

B. Encryption

C. Security awareness training

D. Access control lists

Answer: **D**. Access control lists.

Access control lists are an example of a **preventive control**. Preventive controls are designed to prevent an attacker from gaining unauthorized access to an organization's assets. Intrusion detection systems are an example of a **detective control**, encryption is an example of a **corrective control**, and security awareness training is an example of a **deterrent control**.

39. Which of the following is the best approach to managing third-party risks?

    A. Conducting background checks on all third-party vendors

    B. Requiring all third-party vendors to sign a non-disclosure agreement

    C. Conducting a risk assessment of the third-party vendors

    D. Monitoring all third-party vendors continuously

    Answer: **C**. Conducting a risk assessment of the third-party vendors.

    Conducting a risk assessment of third-party vendors is the best approach to managing **third-party risks**. Third-party risks are the risks that arise when an organization engages with a third-party vendor to provide goods or services. Conducting a risk assessment of the third-party vendor can help identify potential risks associated with the relationship and develop appropriate controls to mitigate those risks. Conducting background checks and requiring non-disclosure agreements are important, but they are not sufficient to manage third-party risks. Monitoring all third-party vendors continuously may be impractical and expensive.

40. Which of the following is the primary purpose of security awareness training?

    A. To develop a security-conscious culture across the organization

    B. To ensure that all employees have a good understanding of information security risks

    C. To ensure that all employees can identify and respond to security incidents

    D. To ensure that all employees are trained in the use of security tools and technologies

    Answer: **A**. To develop a security-conscious culture across the organization.

    The primary purpose of **security awareness training** is to develop a culture of security. Security awareness training is designed to educate employees on the importance of information security, their role in maintaining information security, and the policies and procedures that they must follow to protect information assets. Ensuring that all employees have a good understanding of information security risks, identifying and responding to security incidents, and training employees in the use of security tools and technologies are all important objectives of an organization's information security program, but they are not the primary purpose of security awareness training.

41. Which of the following is an example of a corrective control?

    A. Firewall

    B. Antivirus software

    C. Backup and recovery procedures

    D. Intrusion detection system

Answer: **C**. Backup and recovery procedures.

Backup and recovery procedures are an example of a **corrective control**. Corrective controls are designed to correct or mitigate the effects of an incident or attack after it has occurred. Firewalls and antivirus software are examples of **preventive controls**, whereas an intrusion detection system is an example of a **detective control**.

42. Which of the following is the primary objective of a vulnerability scan?

    A. To identify vulnerabilities in an organization's information systems

    B. To exploit vulnerabilities to test the effectiveness of security controls

    C. To provide a comprehensive assessment of an organization's information security posture

    D. To simulate a real-world attack on an organization's information systems

    Answer: **A**. To identify vulnerabilities in an organization's information systems.

    The primary objective of a **vulnerability scan** is to identify vulnerabilities in an organization's information systems. A vulnerability scan is a tool that scans an organization's information systems to identify potential vulnerabilities that could be exploited by an attacker. Exploiting vulnerabilities to test the effectiveness of security controls is the objective of a penetration test, providing a comprehensive assessment of an organization's information security posture is the objective of a security audit, and simulating a real-world attack on an organization's information systems is the objective of a red team exercise.

43. Which of the following is the best approach to ensure the integrity of data?

    A. Implementing access controls

    B. Implementing encryption

    C. Implementing backup and recovery procedures

    D. Implementing change management procedures

    Answer: **D**. Implementing change management procedures.

    Implementing **change management** procedures is the best approach in this scenario to ensure the integrity of data. Change management is the process of controlling changes to an organization's information systems to minimize the risk of adverse effects on information assets. Implementing access controls and encryption are important steps in ensuring the confidentiality of data, and implementing backup and recovery procedures is important for availability, but none of these controls specifically addresses the integrity of data. Change management procedures can help ensure that changes to data are made in a controlled and authorized manner, minimizing the risk of unauthorized or malicious changes that could compromise the integrity of the data.

44. Which of the following is the primary goal of an information security program?

    A. To eliminate all information security risks

    B. To reduce information security risks to an acceptable level

    C. To eliminate all information security incidents

    D. To minimize the cost of information security controls

    Answer: **B**. To reduce information security risks to an acceptable level.

    The primary goal of an **information security program** is to reduce information security risks to an acceptable level. It is not feasible or practical to eliminate all information security risks or incidents, and minimizing the cost of information security controls is not the primary goal of an information security program.

45. Which of the following is a key component of a security governance framework?

    A. Conducting vulnerability assessments

    B. Conducting security awareness training

    C. Developing security policies and procedures

    D. Installing firewalls and antivirus software

    Answer: **C**. Developing security policies and procedures.

    Developing security policies and procedures is a key component of a **security governance** framework. Security governance refers to the policies, procedures, and controls that are implemented to ensure the effective management of an organization's information security program. Conducting vulnerability assessments, conducting security awareness training, and installing firewalls and antivirus software are important components of an information security program, but they are not specifically related to security governance.

46. Which of the following is the best approach to managing information security risks associated with mobile devices?

    A. Implementing a **mobile device management** (**MDM**) solution

    B. Prohibiting the use of mobile devices

    C. Providing security awareness training to mobile device users

    D. Implementing strong password policies for mobile devices

    Answer: **A**. Implementing a **mobile device management** (**MDM**) solution.

    Implementing an MDM solution is the best approach to managing **information security risks** associated with mobile devices. MDM solutions can help organizations manage mobile devices and ensure that they are configured and used securely. Prohibiting the use of mobile devices is not a practical solution in today's mobile workforce, providing security awareness training is important but not sufficient, and implementing strong password policies is a good practice but does not address all the risks associated with mobile devices.

47. Which of the following is the primary goal of an incident response plan?

    A. To prevent security incidents from occurring

    B. To minimize the impact of security incidents when they occur

    C. To detect security incidents as soon as possible

    D. To identify the root cause of security incidents

    Answer: **B**. To minimize the impact of security incidents when they occur.

    The primary goal of an **incident response plan** is to minimize the impact of security incidents when they occur. Incident response plans provide a framework for responding to security incidents in a timely and effective manner to minimize the impact on the organization's operations, reputation, and assets. While preventing security incidents is a desirable objective, it is not always possible, and incident response plans are designed to manage incidents when they occur. Detecting security incidents as soon as possible and identifying the root cause of incidents are important objectives of an incident response plan, but they are not the primary goal.

48. Which of the following is the best approach to ensure that technical security controls are effective?

    A. Conducting vulnerability assessments

    B. Conducting security awareness training

    C. Conducting security audits

    D. Conducting penetration tests

    Answer: **D**. Conducting penetration tests.

    Conducting penetration tests is the best approach to ensure that **technical security controls are effective**. Penetration tests involve attempting to exploit vulnerabilities in an organization's information systems to determine whether existing security controls are effective in detecting and preventing attacks. Conducting vulnerability assessments and security audits are important components of an information security program, but they do not specifically test the effectiveness of technical security controls. Security awareness training is important to educate employees about security policies and procedures, but it does not specifically test the effectiveness of technical security controls.

49. Which of the following is the best approach to ensure that data backups are available and can be restored in the event of a disaster?

    A. Performing periodic backups and storing them offsite

    B. Storing backup tapes in a secure location

    C. Testing the backup and recovery procedures periodically

    D. Implementing a high-availability solution

    Answer: **C**. Testing the backup and recovery procedures periodically.

    Testing the backup and recovery procedures regularly is the best approach to ensure that **data backups** are available and can be restored in the event of a disaster. Performing periodic backups and storing them offsite, storing backup tapes in a secure location, and implementing a high-availability solution are important components of a disaster recovery plan, but testing the backup and recovery procedures regularly is essential to ensure that the backups can be restored when needed.

50. Which of the following is a key component of a security awareness program?

    A. Conducting vulnerability assessments

    B. Conducting security audits

    C. Developing security policies and procedures

    D. Conducting security awareness training

    Answer: **D**. Conducting security awareness training.

    Conducting security awareness training is a key component of a **security awareness** program. Security awareness programs are designed to educate employees and other stakeholders about security risks and how to protect against them. Conducting vulnerability assessments and security audits are important components of an information security program, but they are not specifically related to security awareness. Developing security policies and procedures is an important component of a security governance framework, but it is not specifically related to security awareness.

51. Which of the following encryption methods uses a single secret key to both encrypt and decrypt data?

    A. Public key encryption

    B. Asymmetric encryption

    C. Symmetric encryption

    D. Hybrid encryption

Answer: **C**. Symmetric encryption.

Symmetric encryption uses a single secret key to both encrypt and decrypt data. This means that the same key is used by both the sender and receiver to encrypt and decrypt the data. **Public key encryption** and **asymmetric encryption** use different keys for encryption and decryption, while **hybrid encryption** uses a combination of both the symmetric and asymmetric encryption methods.

52. Which of the following encryption standards is commonly used for securing wireless networks?

    A. AES

    B. DES

    C. WEP

    D. SSL

    Answer: **C**. WEP.

    Wired Equivalent Privacy (**WEP**) is a commonly used encryption standard for securing wireless networks. However, it is considered a weak encryption standard and is vulnerable to several attacks. **Advanced Encryption Standard** (**AES**) is a stronger encryption standard and is commonly used for securing data in transit and at rest. **Data Encryption Standard** (**DES**) is an older encryption standard that is no longer considered secure. **Secure Sockets Layer** (**SSL**) is a protocol that's used to secure data transmission over the internet, but it is not an encryption standard.

53. Which of the following is a key component of a security risk assessment?

    A. Developing security policies and procedures

    B. Conducting vulnerability assessments

    C. Implementing security controls

    D. Conducting security awareness training

    Answer: **B**. Conducting vulnerability assessments.

    Conducting vulnerability assessments is a key component of a **security risk assessment**. A security risk assessment involves identifying and evaluating information security risks and developing strategies to mitigate or manage them. Conducting vulnerability assessments is an important part of identifying and evaluating information security risks. Developing security policies and procedures, implementing security controls, and conducting security awareness training are all important components of an information security program, but they are not specifically related to security risk assessments.

54. Which of the following is the best approach to protect against social engineering attacks?

    A. Implementing access controls to restrict access to sensitive information

    B. Implementing technical controls to prevent unauthorized access to systems and data

    C. Providing security awareness training to employees

    D. Conducting background checks on employees

    Answer: **C**. Providing security awareness training to employees.

    Providing security awareness training to employees is the best approach to protect against **social engineering attacks**. Social engineering attacks rely on human behavior and psychology to trick people into divulging sensitive information or performing unauthorized actions. Providing security awareness training to employees can help them recognize and resist social engineering attacks. Implementing access controls and technical controls are important components of an information security program, but they do not specifically address social engineering attacks. Conducting background checks is important to vet employees and identify any red flags, but it does not specifically protect against social engineering attacks.

55. Which of the following is a key feature of a hashing algorithm?

    A. It uses a secret key to encrypt data

    B. It is a reversible process

    C. It produces a fixed-length output

    D. It is commonly used for encrypting data in transit

    Answer: **C**. It produces a fixed-length output.

    A key feature of a **hashing algorithm** is that it produces a fixed-length output, regardless of the size of the input data. Hashing algorithms are one-way functions, meaning that they cannot be reversed to obtain the original input data. They are commonly used for verifying the integrity of data, as even a small change to the input data will result in a completely different hash value. Hashing algorithms do not use a secret key to encrypt data, and they are not commonly used for encrypting data in transit, which is typically done using encryption algorithms.

56. Which of the following is the best approach to ensure that security controls are aligned with business objectives?

    A. Conducting a risk assessment

    B. Conducting a gap analysis

    C. Developing a security policy

    D. Developing a business continuity plan

Answer: **B**. Conducting a gap analysis.

Conducting a **gap analysis** is the best approach to ensure that security controls are aligned with business objectives. A gap analysis involves comparing an organization's current security controls and practices to its desired state or other standards, identifying areas where gaps exist, and developing plans to address those gaps. Conducting a risk assessment is important to identify and evaluate information security risks, but it does not specifically ensure that security controls are aligned with business objectives. Developing a security policy and a business continuity plan are important components of a security governance framework, but they do not specifically address whether security controls are aligned with business objectives.

57. Which of the following is a common threat to wireless networks?

    A. Spoofing attacks

    B. SQL injection attacks

    C. Social engineering attacks

    D. Cross-site scripting attacks

    Answer: **A**. Spoofing attacks.

    Spoofing attacks, in which an attacker impersonates a legitimate user or device on a **wireless network**, are a common threat to wireless networks. SQL injection attacks, social engineering attacks, and cross-site scripting attacks are all types of attacks that can be launched against web applications or databases, but they are not specifically related to wireless networks.

58. Which of the following is the best approach to secure a web application against injection attacks?

    A. Implementing access controls to restrict access to sensitive information

    B. Encrypting all data in transit and at rest

    C. Validating all user input

    D. Conducting regular security audits

    Answer: **C**. Validating all user input.

    Validating all user input to prevent injection attacks, such as SQL injection attacks and cross-site scripting attacks, is the best approach to secure a web application. Implementing access controls and encrypting data are important components of an information security program, but they do not specifically address web application security. Conducting regular security audits is important to identify and address vulnerabilities and weaknesses in a web application, but it is not specifically related to preventing injection attacks.

59. Which of the following is the best approach to ensure that a third-party vendor is meeting security requirements?

    A. Conducting regular security audits of the vendor

    B. Requiring the vendor to sign a confidentiality agreement

    C. Conducting a risk assessment of the vendor

    D. Reviewing the vendor's security certifications and attestations

    Answer: **D**. Reviewing the vendor's security certifications and attestations.

    Reviewing the vendor's security certifications and attestations is the best approach to ensure that a third-party vendor is meeting **security requirements**. Security certifications and attestations, such as SOC 2 or ISO 27001, demonstrate that the vendor has implemented appropriate security controls and practices. Conducting regular security audits of the vendor and conducting a risk assessment of the vendor are also important to ensure that the vendor is meeting security requirements, but reviewing the vendor's security certifications and attestations provides a higher level of assurance. Requiring the vendor to sign a confidentiality agreement is important to protect sensitive information, but it does not specifically address security requirements.

60. Which of the following is a key component of a security governance framework?

    A. Implementing technical controls to protect against threats

    B. Developing security policies and procedures

    C. Conducting security awareness training

    D. Monitoring security events and incidents

    Answer: **B**. Developing security policies and procedures.

    Developing security policies and procedures is a key component of a **security governance** framework. Security governance involves establishing and maintaining an information security program that aligns with business objectives, regulatory requirements, and industry best practices. Developing security policies and procedures provides a foundation for a comprehensive information security program that addresses risks and threats, establishes appropriate controls, and defines roles and responsibilities. Implementing technical controls, conducting security awareness training, and monitoring security events and incidents are all important components of an information security program, but they are not specifically related to security governance.

61. Which of the following is the best approach to prevent data leakage through email?

    A. Implementing **data loss prevention** (**DLP**) software on all email systems

    B. Encrypting all email messages containing sensitive information

    C.   Restricting access to email systems to authorized personnel only

    D. Conducting regular security audits of email systems

Answer: **A**. Implementing **data loss prevention (DLP)** software on all email systems.

Implementing DLP software on all email systems is the best approach to prevent **data leakage** through email. DLP software can identify sensitive information in email messages and prevent it from being sent outside the organization or to unauthorized individuals. Encrypting email messages and restricting access to email systems are also important components of an information security program, but they do not specifically address data leakage through email. Conducting regular security audits of email systems is important to identify and address vulnerabilities and weaknesses, but it is not specifically related to preventing data leakage through email.

62. Which of the following is a key component of an incident response plan?

    A. Identifying critical business processes and assets

    B. Developing communication procedures

    C. Conducting regular security awareness training

    D. Installing intrusion detection and prevention systems

Answer: **B**. Developing communication procedures.

Developing communication procedures is a key component of an **incident response plan**. Communication is critical during an incident to ensure that all stakeholders are informed and can coordinate their response. Identifying critical business processes and assets is important to prioritize incident response efforts, but it is not specifically related to communication procedures. Conducting regular security awareness training and installing intrusion detection and prevention systems are important components of an information security program, but they are not specifically related to incident response.

63. Which of the following is a key advantage of using cloud computing?

    A. Reduced scalability

    B. Lower security risks

    C. Increased capital expenditures

    D. On-demand usage pricing

Answer: **D**. On-demand usage pricing.

On-demand usage pricing is a key advantage of using **cloud computing** as it allows organizations to only pay for the resources they use, rather than investing in and maintaining their infrastructure. Cloud computing can also provide increased scalability, but not reduced scalability, as well as potential security risks and reduced capital expenditure.

64. Which of the following cloud deployment models involves sharing computing resources among multiple organizations with similar requirements?

    A. Private cloud

    B. Public cloud

    C. Community cloud

    D. Hybrid cloud

    Answer: **C**. Community cloud.

    A community cloud involves sharing computing resources among multiple organizations that have common concerns, such as regulatory compliance or security requirements. The private cloud is dedicated to a single organization, while the public cloud is open to the general public. A hybrid cloud involves a combination of two or more cloud deployment models, typically private and public clouds.

65. Which of the following is a key feature of a digital signature?

    A. It uses a secret key to encrypt data

    B. It verifies the integrity of the data

    C. It is always visible to the recipient

    D. It is the same as an electronic signature

    Answer: **B**. It verifies the integrity of the data.

    A key feature of a **digital signature** is that it verifies the integrity of the data, ensuring that the data has not been tampered with since it was signed. Digital signatures use a combination of public and private keys to provide this verification, but they do not encrypt the data. Digital signatures are not always visible to the recipient and they are not the same as electronic signatures, which do not provide the same level of security.

66. Which of the following is a commonly used digital signature standard?

    A. RSA

    B. AES

    C. SSL

    D. FTP

    Answer: **A**. RSA.

    Rivest-Shamir-Adleman (RSA) is a commonly used digital signature standard, as well as a public key cryptography algorithm. RSA uses a combination of public and private keys to sign and verify digital signatures. **Advanced Encryption Standard (AES)** and **Secure Sockets**

Layer (**SSL**) are encryption standards, while **File Transfer Protocol** (**FTP**) is a protocol used for transferring files over the internet.

67. Which of the following is the best approach to managing the risks associated with remote access?

    A. Restricting all remote access to a VPN

    B. Implementing MFA for all remote access

    C. Limiting the use of personal devices for remote access

    D. Conducting regular security audits of remote access systems

    Answer: **A**. Restricting all remote access to a VPN.

    Restricting all **remote access** to a VPN is the best approach to managing the risks associated with remote access. A VPN provides a secure, encrypted connection between a remote device and the organization's network, helping prevent unauthorized access and data breaches. Implementing MFA and limiting the use of personal devices are also important components of an information security program, but they do not specifically address the risks associated with remote access. Conducting regular security audits of remote access systems is important to identify and address vulnerabilities and weaknesses, but it is not specifically related to managing the risks associated with remote access.

68. Which of the following is the best approach to ensure that security controls are implemented consistently across the organization?

    A. Conducting regular security awareness training for all employees

    B. Developing and enforcing security policies and procedures

    C. Conducting regular security audits of all systems

    D. Implementing intrusion detection and prevention systems on all systems

    Answer: **B**. Developing and enforcing security policies and procedures.

    Developing and enforcing security policies and procedures is the best approach to ensure that **security controls** are implemented consistently across the organization. Security policies and procedures define the rules and guidelines for how security controls should be implemented and enforced, and they provide a framework for ensuring consistency across the organization. Conducting regular security awareness training, conducting security audits, and implementing intrusion detection and prevention systems are all important components of an information security program, but they are not specifically related to ensuring consistency in security controls.

69. Which of the following is a common vulnerability associated with mobile devices?

    A. Cross-site scripting attacks

    B. Man-in-the-middle attacks

    C. Rooting or jailbreaking

    D. SQL injection attacks

    Answer: **C**. Rooting or jailbreaking.

    Rooting or **jailbreaking** is a common **vulnerability** associated with mobile devices. Rooting or jailbreaking is the process of removing the restrictions imposed by the device manufacturer or operating system, allowing users to install apps and modify the device's operating system. However, this also removes the security controls and opens the device to potential security risks. Cross-site scripting attacks, man-in-the-middle attacks, and SQL injection attacks are all types of attacks that can target mobile devices, but they are not vulnerabilities specifically associated with mobile devices.

70. Which of the following is the best approach to ensure the confidentiality and integrity of data in transit?

    A. Implementing access controls on all data storage systems

    B. Implementing DLP software on all email systems

    C. Encrypting all data transmissions using secure protocols

    D. Conducting regular vulnerability assessments of all systems

    Answer: **C**. Encrypting all data transmissions using secure protocols.

    Encrypting all data transmissions using secure protocols is the best approach to ensure the **confidentiality** and **integrity** of data in transit. Encryption protects data from unauthorized access and modification while it is being transmitted over a network. Implementing access controls on data storage systems and conducting regular vulnerability assessments are important components of an information security program, but they do not specifically address the security of data in transit. Implementing DLP software on email systems is important to prevent data leakage through email, but it is not specifically related to ensuring the confidentiality and integrity of data in transit.

71. Which of the following is a common vulnerability associated with cloud computing?

    A. Denial-of-service attacks

    B. Cross-site scripting attacks

    C. SQL injection attacks

    D. Insufficient access controls

Answer: **D**. Insufficient access controls.

Insufficient access controls are a common **vulnerability** associated with **cloud computing**. Cloud computing environments are typically complex and dynamic, with multiple users, applications, and data sources accessing shared resources. Without effective access controls, unauthorized users can access sensitive data or resources. Denial-of-service attacks, cross-site scripting attacks, and SQL injection attacks are all types of attacks that can target cloud computing environments, but they are not vulnerabilities specifically associated with cloud computing.

72. Which of the following is a key component of a risk management program?

    A. Developing and enforcing security policies and procedures

    B. Conducting regular security awareness training

    C. Conducting risk assessments

    D. Implementing access controls on all systems

    Answer: **C**. Conducting risk assessments.

    Conducting risk assessments is a key component of a **risk management** program. Risk assessments identify and prioritize potential risks and vulnerabilities to the organization's information assets, allowing for effective mitigation strategies to be developed and implemented. Developing and enforcing security policies and procedures, conducting security awareness training, and implementing access controls are all important components of an information security program, but they are not specifically related to risk management.

73. Which of the following is a key component of a business continuity plan?

    A. Developing communication procedures

    B. Conducting regular security audits

    C. Implementing intrusion detection and prevention systems

    D. Developing and testing recovery procedures

    Answer: **D**. Developing and testing recovery procedures.

    Developing and testing recovery procedures is a key component of a **business continuity plan**. Recovery procedures define the steps that should be taken to restore critical business processes and systems in the event of a disruption or disaster, and testing ensures that the procedures are effective and can be executed quickly and efficiently. Developing communication procedures, conducting security audits, and implementing intrusion detection and prevention systems are important components of an information security program, but they are not specifically related to business continuity planning.

74. Which of the following is a common vulnerability associated with social engineering attacks?

   A. Cross-site scripting attacks

   B. SQL injection attacks

   C. Insider threats

   D. Human error

   Answer: **D**. Human error.

   Social engineering attacks rely on human error or manipulation to gain access to sensitive information or resources. This can include techniques such as phishing, pretexting, or baiting, which are designed to trick individuals into revealing confidential information or granting unauthorized access. While cross-site scripting attacks, SQL injection attacks, and insider threats are all potential vulnerabilities in an organization's security, they are not specific to social engineering attacks.

75. Your organization is considering outsourcing some IT services to a third-party vendor. What is the most critical control to ensure that the vendor's security practices meet your organization's standards?

   A. Signing a non-disclosure agreement with the vendor

   B. Conducting an annual third-party risk assessment

   C. Ensuring that the vendor has ISO 27001 certification

   D. Performing a background check on the vendor's employees

   Answer: **B**. Conducting an annual third-party risk assessment.

   Conducting an annual third-party risk assessment is the most critical control to ensure that the vendor's **security practices** meet your organization's standards. It will help identify and evaluate the risks associated with outsourcing IT services, including any vulnerabilities or gaps in the vendor's security controls. Signing a non-disclosure agreement with the vendor is important, but does not ensure that the vendor's security practices meet your organization's standards. ISO 27001 certification is a good indicator of a vendor's security posture, but it does not guarantee that they meet your specific security requirements. Performing a background check on the vendor's employees is important, but it is not the most critical control to ensure that the vendor's security practices meet your organization's standards.

76. Your organization is implementing a new **identity and access management** (**IAM**) solution. What is the most important feature that should be included in the solution?

    A. SSO capabilities

    B. Password complexity rules

    C. MFA

    D. RBAC

    Answer: **D**. RBAC.

    RBAC is the most important feature that should be included in an IAM solution. RBAC ensures that users are granted access based on their job functions, which minimizes the risk of unauthorized access to sensitive information. SSO capabilities are important, but they are not the most important feature of an IAM solution. Password complexity rules and MFA are important for securing user credentials, but they are not the most important feature of an IAM solution.

77. Your organization has just experienced a data breach that resulted in the loss of sensitive customer information. What is the most critical action that should be taken immediately?

    A. Notify the affected customers and offer them identity theft protection services

    B. Conduct a forensic investigation to determine the cause and extent of the breach

    C. Change all user passwords and implement MFA

    D. File a report with the appropriate regulatory authorities

    Answer: **B**. Conduct a forensic investigation to determine the cause and extent of the breach.

    Conducting a forensic investigation to determine the cause and extent of the breach is the most critical action that should be taken immediately. It will help identify the root cause of the breach and any vulnerabilities or gaps in the organization's security controls. Notifying the affected customers and offering them identity theft protection services is important, but it should not be done until the forensic investigation is complete. Changing all user passwords and implementing MFA is important, but it should be done after the root cause of the breach has been identified and addressed. Filing a report with the appropriate regulatory authorities is important, but it is not the most critical action to take immediately.

78. Your organization is implementing a new business intelligence system that will be used to store and analyze large amounts of sensitive data. What is the most important control to ensure the confidentiality of the data?

    A. Implementing DLP software

    B. Encrypting the data at rest and in transit

    C. Implementing access controls and audit trails

    D. Conducting regular vulnerability scans and penetration tests

Answer: **B**. Encrypting the data at rest and in transit.

Encrypting the data at rest and in transit is the most important control to ensure the **confidentiality** of the data. This will ensure that even if the data is accessed by an unauthorized user, they will not be able to read it. Implementing DLP software, access controls, and audit trails are important, but they do not ensure the confidentiality of the data. Conducting regular vulnerability scans and penetration tests is important, but it is not the most important control to ensure the confidentiality of the data.

79. Your organization is implementing a new **bring your own device (BYOD)** policy. What is the most important control to ensure the security of the devices?

   A.   Requiring users to install antivirus software on their devices

   B.   Implementing MDM software

   C.   Requiring users to use strong passwords and enable device encryption

   D.   Limiting the types of applications that can be installed on the devices

Answer: **B**. Implementing MDM software.

Implementing MDM software is the most important control to ensure the security of BYOD devices. MDM software can be used to enforce security policies, remotely wipe devices, and manage application installations. Requiring users to install antivirus software, requiring users to use strong passwords and enable device encryption, and limiting the types of applications that can be installed are important, but they do not ensure the security of BYOD devices to the same extent as MDM software.

80. Your organization is implementing a new e-commerce website. What is the most important control to ensure the integrity of customer transactions?

   A.   Implementing SSL encryption

   B.   Conducting regular vulnerability scans and penetration tests

   C.   Implementing access controls and audit trails

   D.   Implementing input validation and output sanitization

Answer: **D**. Implementing input validation and output sanitization.

Implementing input validation and output sanitization is the most important control to ensure the **integrity** of customer transactions. This will prevent attackers from injecting malicious code into the website, which could lead to the manipulation of customer transactions. Implementing SSL encryption is important, but it does not ensure the integrity of customer transactions to the same extent as input validation and output sanitization. Conducting regular vulnerability scans and penetration tests and implementing access controls and audit trails are important, but they do not ensure the integrity of customer transactions.

81. Your organization is implementing a new cloud-based HR system. What is the most important control to ensure the confidentiality of employee data?

    A. Implementing access controls and audit trails

    B. Conducting regular vulnerability scans and penetration tests

    C. Requiring the use of strong passwords and two-factor authentication

    D. Implementing data encryption for data at rest and in transit

    Answer: **D**. Implementing data encryption for data at rest and in transit.

    Implementing data encryption for data at rest and in transit is the most important control to ensure the **confidentiality** of employee data in a cloud-based HR system. Access controls and audit trails are important, but they do not ensure the confidentiality of employee data to the same extent as encryption. Conducting regular vulnerability scans and penetration tests is important, but it is not the most important control to ensure the confidentiality of employee data. Requiring the use of strong passwords and two-factor authentication will not prevent the confidentiality of data.

82. A company is planning to migrate its IT infrastructure to the cloud. As an information security manager, what is the *most* important consideration to address before the migration?

    A. The data classification of the company's sensitive information

    B. The reliability of the cloud provider's infrastructure

    C. The availability of security controls in the cloud environment

    D. The compatibility of the company's existing software with the cloud environment

    Answer: **A**. The data classification of the company's sensitive information.

    Before migrating to the cloud, it is crucial to determine the data classification of the company's sensitive information. This will help determine the appropriate level of **security controls** needed to protect the data in the cloud environment.

83. A company has implemented a new security control to prevent unauthorized access to its network. Which of the following is the *best* way to evaluate the effectiveness of the control?

    A. Conducting a penetration test

    B. Reviewing the control logs

    C. Conducting a risk assessment

    D. Performing a vulnerability scan

Answer: **B**. Reviewing the control logs.

Reviewing the control logs is the best way to evaluate the effectiveness of the **security control**. This will help determine whether the control is working as intended and whether any unauthorized access attempts were made.

A is incorrect because conducting a penetration test is a way to test the overall security posture of the company's network, not just the effectiveness of a single control.

C is incorrect because conducting a risk assessment is a way to identify and evaluate risks to the company's assets, not to evaluate the effectiveness of a single control.

D is incorrect because performing a vulnerability scan is a way to identify vulnerabilities in the company's network, not to evaluate the effectiveness of a single control.

84. A company is implementing a new data classification policy. Which of the following is the *best* way to ensure that the policy is effective?

    A. Conducting security awareness training for employees

    B. Conducting periodic audits of the policy

    C. Conducting a penetration test on the company's network

    D. Conducting a vulnerability scan on the company's network

    Answer: **A**. Conducting security awareness training for employees.

    Conducting security awareness training for employees is the best way to ensure that the new **data classification policy** is effective. This will help ensure that employees understand the policy and can implement it properly.

    B is incorrect because while conducting periodic audits of the policy is important, it is *not* the best way to ensure the policy is effective.

    C and *D* are incorrect because conducting a penetration test or vulnerability scan is *not* a relevant way to ensure the effectiveness of a data classification policy.

85. A manufacturing company is planning to implement a new **enterprise resource planning (ERP)** system. Which of the following is the most important factor to consider when assessing the risk associated with this implementation?

    A. The cost of the new system

    B. The potential impact on business processes

    C. The number of users who will have access to the system

    D. The reputation of the vendor providing the system

    Answer: **B**. The potential impact on business processes.

    When implementing a new ERP system, the potential impact on business processes is a critical factor to consider when assessing the **risk** associated with the implementation. The cost of the system, the number of users who will have access to it, and the reputation of the vendor are all important factors, but they are secondary to the potential impact on business processes.

86. A financial services company is planning to migrate its data center to the cloud. Which of the following is the primary risk associated with this migration?

    A. Increased complexity of the infrastructure

    B. Decreased control over the infrastructure

    C. Increased cost of maintaining the infrastructure

    D. Decreased availability of the infrastructure

    Answer: **B**. Decreased control over the infrastructure.

    When migrating a data center to the **cloud**, the primary risk associated with the migration is decreased control over the infrastructure. This is because the infrastructure is owned and managed by the cloud service provider, which means that the organization will have less visibility and control over the infrastructure.

87. A healthcare organization is implementing a new electronic health record system. Which of the following is the most important factor to consider when assessing the risk associated with this implementation?

    A. The potential impact on patient privacy

    B. The cost of the new system

    C. The number of users who will have access to the system

    D. The reputation of the vendor providing the system

    Answer: **A**. The potential impact on patient privacy.

    When implementing a new electronic health record system, the most important factor to consider when assessing the **risk** associated with the implementation is the potential impact on patient privacy. Healthcare organizations have a legal and ethical obligation to protect patient privacy, so any system that stores and processes patient information must be secure and compliant with relevant regulations.

88. Which of the following is the most important control to implement to mitigate the risk of unauthorized access to the point-of-sale system?

    A. Strong passwords for user accounts

    B. Encryption of payment card data

    C. RBAC

    D. Regular vulnerability scans of the system

Answer: **C**. RBAC.

When implementing a new point-of-sale system, the most important control to implement to mitigate the risk of **unauthorized access** to the system is RBAC. This control ensures that users are granted access only to the data and functionality that they need to perform their job duties.

89. A government agency is implementing a new system to store and process classified information. Which of the following is the most important factor to consider when assessing the risk associated with this implementation?

    A.   The potential impact on national security

    B.   The cost of the new system

    C.   The number of users who will have access to the system

    D.   The reputation of the vendor providing the system

    Answer: **A**. The potential impact on national security.

    When implementing a new system to store and process **classified information**, the most important factor to consider when assessing the risk associated with the implementation is the potential impact on national security. This is because classified information being compromised can have serious consequences for national security.

90. A university is implementing a new learning management system. Which of the following is the most important control to implement to mitigate the risk of data breaches?

    A.   Encryption of data in transit

    B.   Strong passwords for user accounts

    C.   Regular backups of the system

    D.   MFA for user accounts

    Answer: **A**. Encryption of data in transit.

    When implementing a new learning management system, the most important control to implement to mitigate the risk of **data breaches** is the encryption of data in transit. This control ensures that data is protected as it is transmitted over the network.

91. A financial institution is implementing a new online banking platform. Which of the following is the most important control to implement to mitigate the risk of unauthorized transactions?

    A.   Encryption of data at rest

    B.   MFA for user accounts

    C.   Regular security awareness training for employees

    D.   Regular penetration testing of the system

Answer: **B**. MFA for user accounts.

When implementing a new online banking platform, the most important control to implement to mitigate the risk of **unauthorized transactions** is MFA for user accounts. This control ensures that users are required to provide multiple pieces of evidence to prove their identity before they are allowed to access their accounts.

92. An e-commerce company is implementing a new payment processing system. Which of the following is the most important control to implement to mitigate the risk of payment card fraud?

    A. Encryption of payment card data

    B. Strong passwords for user accounts

    C. Regular backups of the system

    D. Regular vulnerability scans of the system

    Answer: **A**. Encryption of payment card data.

    When implementing a new payment processing system, the most important control to implement to mitigate the risk of **payment card fraud** is the encryption of payment card data. This control ensures that payment card data is protected from unauthorized access.

93. A healthcare provider is implementing a new electronic medical records system. Which of the following is the most important control to implement to mitigate the risk of unauthorized access to patient data?

    A. RBAC

    B. Regular backups of the system

    C. Encryption of data at rest

    D. MFA for user accounts

    Answer: **A**. RBAC.

    When implementing a new electronic medical records system, the most important control to implement to mitigate the risk of **unauthorized access** to patient data is RBAC. This control ensures that users are granted access only to the data and functionality that they need to perform their job duties.

94. A manufacturing company is implementing a new supply chain management system. Which of the following is the most important control to implement to mitigate the risk of supply chain disruptions?

    A. A vendor risk management program

    B. Regular backups of the system

    C. Regular security awareness training for employees

    D. Regular vulnerability scans of the system

Answer: **A**. A vendor risk management program.

When implementing a new supply chain management system, the most important control to implement to mitigate the risk of **supply chain disruptions** is a vendor risk management program. This control ensures that the company evaluates the risk associated with its vendors and takes steps to mitigate those risks.

95. A financial institution is implementing a new trading platform. Which of the following is the most important control to implement to mitigate the risk of unauthorized trading activity?

    A. RBAC

    B. Regular backups of the system

    C. Encryption of data at rest

    D. MFA for user accounts

    Answer: **A**. RBAC.

    When implementing a new trading platform, the most important control to implement to mitigate the risk of **unauthorized trading activity** is RBAC. This control ensures that users are granted access only to the data and functionality that they need to perform their job duties.

96. A retail company is implementing a new e-commerce website. Which of the following is the most important control to implement to mitigate the risk of credit card fraud?

    A. Encryption of credit card data

    B. Regular backups of the system

    C. MFA for user accounts

    D. Regular vulnerability scans of the system

    Answer: **A**. Encryption of credit card data.

    When implementing a new e-commerce website, the most important control to implement to mitigate the risk of **credit card fraud** is the encryption of credit card data. This control ensures that credit card data is protected from unauthorized access.

97. An educational institution is implementing a new learning management system. Which of the following is the most important control to implement to mitigate the risk of unauthorized access to student data?

    A. RBAC

    B. Regular backups of the system

    C. Encryption of data at rest

    D. MFA for user accounts

Answer: **A**. RBAC.

When implementing a new learning management system, the most important control to implement to mitigate the risk of **unauthorized access** to student data is RBAC. This control ensures that users are granted access only to the data and functionality that they need to perform their job duties.

98. A healthcare organization is implementing a new electronic health record system. Which of the following is the most important control to implement to mitigate the risk of data breaches?

   A.   Encryption of data at rest

   B.   Regular vulnerability scans of the system

   C.   MFA for user accounts

   D.   Regular backups of the system

   Answer: **A**. Encryption of data at rest.

When implementing a new electronic health record system, the most important control to implement to mitigate the risk of **data breaches** is the encryption of data at rest. This control ensures that sensitive data is protected from unauthorized access in case of a breach.

99. A transportation company is implementing a new dispatch system. Which of the following is the most important control to implement to mitigate the risk of system downtime?

   A.   Redundant hardware components

   B.   Regular vulnerability scans of the system

   C.   MFA for user accounts

   D.   Regular backups of the system

   Answer: **A**. Redundant hardware components.

When implementing a new dispatch system, the most important control to implement to mitigate the risk of **system downtime** is redundant hardware components. This control ensures that if a component of the system fails, there is another component that can take over its function.

100. A manufacturing company is implementing a new system for tracking inventory. Which of the following is the most important control to implement to mitigate the risk of data loss?

   A.   Regular backups of the system

   B.   Encryption of data at rest

   C.   MFA for user accounts

   D.   RBAC

Answer: **A**. Regular backups of the system.

When implementing a new system for tracking inventory, the most important control to implement to mitigate the risk of **data loss** is regular backups of the system. This control ensures that if data is lost or corrupted, it can be recovered from a backup.

19
Practice Quiz – Part 2

1. Which of the following best describes a threat?

   A. The likelihood of an event occurring

   B. A weakness in a system or process

   C. An event or action that has the potential to cause harm

   D. The impact of an event or action

   Answer: **C**. An event or action that has the potential to cause harm.

   A **threat** is an event or action that has the potential to cause harm to an organization's assets, such as its data, systems, or people. Threats can be intentional, such as a cyberattack or theft, or unintentional, such as a natural disaster or equipment failure.

2. Which of the following best describes a vulnerability?

   A. The likelihood of an event occurring

   B. A weakness in a system or process

   C. An event or action that has the potential to cause harm

   D. The impact of an event or action

   Answer: **B**. A weakness in a system or process.

   A **vulnerability** is a weakness in a system or process that can be exploited by a threat to cause harm to an organization's assets. Vulnerabilities can be due to factors such as outdated software, poor configuration, or human error.

3. Which of the following best describes a risk?

    A. The likelihood of an event occurring

    B. A weakness in a system or process

    C. An event or action that has the potential to cause harm

    D. The impact of an event or action

    Answer: **D**. The impact of an event or action.

    Risk is the potential impact of an event or action on an organization's assets. Risk is often expressed in terms of **likelihood** and **impact**, with *likelihood* referring to the probability of the event occurring and *impact* referring to the magnitude of the consequences if the event were to occur.

4. Which of the following is an example of a risk management strategy?

    A. Implementing a firewall to protect against cyberattacks

    B. Conducting periodic vulnerability assessments

    C. Accepting the risk of a potential event and not taking any action

    D. Identifying potential threats to an organization's assets

    Answer: **A**. Implementing a firewall to protect against cyberattacks.

    Implementing a firewall to protect against cyberattacks is an example of a **risk management strategy**, as it is a control measure designed to mitigate the risk of a potential threat to an organization's assets. Conducting periodic vulnerability assessments and identifying potential threats are **risk assessment** activities, while accepting the risk of a potential event and not taking any action is an example of a **risk acceptance** strategy.

5. Which of the following is a primary goal of a business impact analysis?

    A. To identify and prioritize critical business processes and systems

    B. To assess the likelihood and impact of potential risks

    C. To establish recovery time objectives for critical systems

    D. To identify appropriate risk management strategies

    Answer: **A**. To identify and prioritize critical business processes and systems.

    The primary goal of a **business impact analysis (BIA)** is to identify and prioritize critical business processes and systems based on their importance to an organization's operations. This allows organizations to prioritize their resources and efforts in the event of a disruption or disaster.

6. Which of the following best describes inherent risk?

    A. The level of risk remaining after controls have been implemented

    B. The level of risk prior to implementing any controls

    C. The level of risk associated with a specific event or action

    D. The level of risk associated with a specific vulnerability

    Answer: **B**. The level of risk prior to implementing any controls.

    Inherent risk is the level of risk that exists prior to implementing any controls to mitigate the risk. It represents the level of risk that an organization would face if no controls were in place.

7. Which of the following best describes residual risk?

    A. The level of risk remaining after controls have been implemented

    B. The level of risk prior to implementing any controls

    C. The level of risk associated with a specific event or action

    D. The level of risk associated with a specific vulnerability

    Answer: **A**. The level of risk remaining after controls have been implemented.

    Residual risk is the level of risk that remains after controls have been implemented to mitigate the inherent risk. It represents the level of risk that an organization still faces even after implementing controls to reduce the risk.

8. Which of the following is an example of a risk mitigation strategy?

    A. Implementing a backup generator to ensure power availability during an outage

    B. Conducting a BIA to identify critical systems and processes

    C. Accepting the risk associated with a low-likelihood event

    D. Prioritizing high-risk systems for testing and validation

    Answer: **A**. Implementing a backup generator to ensure power availability during an outage.

    Implementing a backup generator to ensure power availability during an outage is an example of a **risk mitigation** strategy, as it is a control measure designed to reduce the risk associated with a potential event. Conducting a BIA, accepting the risk associated with a low-likelihood event, and prioritizing high-risk systems for testing and validation are all **risk assessment** activities.

9. What is the primary difference between an RPO and an RTO?

    A. RPO measures the maximum amount of data loss an organization can tolerate, while RTO measures the maximum amount of downtime an organization can tolerate

    B. RTO measures the maximum amount of data loss an organization can tolerate, while RPO measures the maximum amount of downtime an organization can tolerate

    C. RPO measures the maximum amount of time it takes to restore data after a disaster, while RTO measures the maximum amount of time it takes to restore operations after a disaster

    D. RTO measures the maximum amount of time it takes to restore data after a disaster, while RPO measures the maximum amount of time it takes to restore operations after a disaster

    Answer: **A**. RPO measures the maximum amount of data loss an organization can tolerate, while RTO measures the maximum amount of downtime an organization can tolerate.

    RPO and RTO are both critical metrics used to measure an organization's **recovery capability** in the event of a disaster.

10. Which of the following would be the preferable **maximum tolerable downtime (MTD)** for a critical system that must be operational 24/7?

    A. 1 hour

    B. 4 hours

    C. 8 hours

    D. 12 hours

    Answer: **A**. 1 hour.

    The MTD is the maximum amount of time that a critical system can be down before the organization faces significant financial or operational consequences. For a critical system that must be operational 24/7, the MTD would be very *low*, typically measured in minutes or hours. Therefore, the correct answer is 1 hour. It should be noted that 1 hour is not an industry standard but is the best answer of all the available options.

11. What should be the **maximum tolerable time to detect (MTTD)** for a critical security incident?

    A. 1 hour

    B. 4 hours

    C. 8 hours

    D. 12 hours

Answer: **A**. 1 hour.

MTTD is the maximum amount of time that an organization can tolerate between the occurrence of a security incident and its detection. For a critical security incident, the MTTD should be as *low* as possible to minimize the impact of the incident. Typically, the MTTD for a critical security incident is measured in minutes or hours, so the correct answer is 1 hour.

12. What is the difference between RTO and MTD?

    A. RTO measures the maximum amount of downtime an organization can tolerate, while MTD measures the maximum amount of time it takes to restore operations after a disaster.

    B. RTO measures the maximum amount of time it takes to restore data after a disaster, while MTD measures the maximum amount of time that a critical system can be down before the organization faces significant financial or operational consequences.

    C. RTO measures the maximum amount of data loss an organization can tolerate, while MTD measures the maximum amount of time that an organization can tolerate between the occurrence of a security incident and its detection.

    D. RTO measures the maximum amount of time it takes to restore operations after a disaster, while MTD measures the maximum amount of time that an organization can tolerate between the occurrence of a security incident and its containment.

    Answer: **B**. RTO measures the maximum amount of time it takes to restore data after a disaster, while MTD measures the maximum amount of time that a critical system can be down before the organization faces significant financial or operational consequences.

    RTO and MTD are both critical metrics used to measure an organization's **recovery capability** in the event of a disaster or outage.

13. What is the difference between RPO and MTTD?

    A. RPO measures the maximum amount of data loss an organization can tolerate, while MTTD measures the maximum amount of time that an organization can tolerate between the occurrence of a security incident and its detection.

    B. RPO measures the maximum amount of time it takes to restore data after a disaster, while MTTD measures the maximum amount of time it takes to restore operations after a disaster.

    C. RPO measures the maximum amount of time that a critical system can be down before the organization faces significant financial or operational consequences, while MTTD measures the maximum amount of time it takes to restore data after a disaster.

    D. RPO measures the maximum amount of time that an organization can tolerate between the occurrence of a security incident and its detection, while MTTD measures the maximum amount of data loss an organization can tolerate.

Answer: **A**. RPO measures the maximum amount of data loss an organization can tolerate, while MTTD measures the maximum amount of time that an organization can tolerate between the occurrence of a security incident and its detection.

RPO measures the maximum amount of data loss an organization can tolerate, while MTTD measures the maximum amount of time that an organization can tolerate between the occurrence of a security incident and its detection. RPO and MTTD are both critical metrics used to measure an organization's **resilience** against data loss and security incidents.

14. What is the primary objective of **Third-Party Risk Management (TPRM)**?

    A.   To manage internal risks within an organization

    B.   To manage external risks posed by third-party vendors or suppliers

    C.   To manage compliance with regulatory requirements

    D.   To manage financial risks associated with procurement

    Answer: **B**. To manage external risks posed by third-party vendors or suppliers.

    The primary objective of TPRM is to manage the **external risks** posed by third-party vendors or suppliers, including risks related to security, data privacy, compliance, and financial stability.

15. Which of the following is *NOT* a key component of TPRM?

    A.   Vendor selection

    B.   Due diligence

    C.   Risk assessment

    D.   Business continuity planning

    Answer: **D**. Business continuity planning.

    While business continuity planning is an important aspect of overall **risk management**, it is *not* a key component of TPRM. The key components of TPRM are vendor selection, due diligence, risk assessment, risk mitigation, and ongoing monitoring and management.

16. What is the primary objective of a **Security Information and Event Management (SIEM)** system?

    A.   To monitor network traffic and detect intrusions

    B.   To analyze and correlate security event data from multiple sources

    C.   To prevent malware infections on endpoints

    D.   To provide real-time visibility into system availability

    Answer: **B**. To analyze and correlate security event data from multiple sources.

    The primary objective of a SIEM system is to analyze and correlate **security event data** from multiple sources, including firewalls, intrusion detection systems, and antivirus software, in order to detect and respond to security threats in real time.

17. Which of the following is *NOT* a key feature of a SIEM system?

    A. Event correlation

    B. User authentication

    C. Log management

    D. Incident response

    Answer: **B**. User authentication.

    While user authentication is an important aspect of overall **security management**, it is *not* a key feature of a SIEM system. The key features of a SIEM system are event correlation, log management, incident response, and real-time security monitoring.

18. What is the difference between a rule-based SIEM and a behavior-based SIEM?

    A. A rule-based SIEM only detects known threats, while a behavior-based SIEM can detect unknown threats.

    B. A rule-based SIEM can only detect events that match predefined rules, while a behavior-based SIEM can detect abnormal patterns of activity.

    C. A rule-based SIEM is less expensive than a behavior-based SIEM.

    D. A behavior-based SIEM requires less maintenance than a rule-based SIEM.

    Answer: **B**. A rule-based SIEM can only detect events that match predefined rules, while a behavior-based SIEM can detect abnormal patterns of activity.

    The main difference between a **rule-based SIEM** and a **behavior-based SIEM** is the way in which they detect security events. A rule-based SIEM uses predefined rules to detect specific security events, while a behavior-based SIEM uses machine learning algorithms to detect abnormal patterns of activity that may indicate a security threat. As a result, a behavior-based SIEM is generally considered to be *more effective* at detecting unknown or emerging threats.

19. What does CMM stand for in the context of information security?

    A. Cybersecurity Maturity Model

    B. Capability Maturity Model

    C. Configuration Management Model

    D. Computerized Maintenance Management

    Answer: **B**. Capability Maturity Model.

    The **Capability Maturity Model** (**CMM**) is a framework that describes the maturity of an organization's processes and procedures. It is often used in the context of software development and information security to measure the **maturity** of an organization's security processes and to identify areas for improvement.

20. Which of the following is *NOT* a level of maturity in the CMM framework?

   A. Level 0 – incomplete

   B. Level 2 – repeatable

   C. Level 4 – managed

   D. Level 6 – optimized

   Answer: **D**. Level 6 – optimized.

   The CMM consists of five levels of maturity, from *Level 0 – incomplete* to *Level 5 – optimizing*. Level 5 is the highest level of maturity in the CMM framework. There is no Level 6 in the CMM framework.

21. Which of the following is a transport layer protocol in the TCP/IP model?

   A. IP

   B. ARP

   C. UDP

   D. ICMP

   Answer: **C**. UDP.

   The **User Datagram Protocol** (**UDP**) is a transport layer protocol in the **TCP/IP model**. UDP is a connectionless protocol that provides a lightweight, low-overhead method to send datagrams between devices on a network.

22. Which layer of the OSI model is responsible for converting data between the format used by the application and the format used by the user?

   A. Physical layer

   B. Data link layer

   C. Presentation layer

   D. Session layer

   Answer: **C**. Presentation layer.

   The **presentation layer** of the **OSI model** is responsible for converting data between the format used by the application and the format used by the user. This layer is responsible for data compression, encryption, and formatting.

23. What is the purpose of a firewall in a network security architecture?

    A. To prevent unauthorized access to the network

    B. To encrypt all network traffic

    C. To provide secure remote access to the network

    D. To monitor network activity for security threats

    Answer: **A**. To prevent unauthorized access to the network.

    A **firewall** is a network security device that monitors and regulates incoming and outgoing **network traffic**. Its main purpose is to prevent unauthorized access to the network by blocking traffic that does not meet specified security criteria.

24. Which of the following types of firewalls operates at the transport layer of the OSI model?

    A. Packet filter firewall

    B. Stateful inspection firewall

    C. Application-level gateway

    D. Circuit-level gateway

    Answer: **B**. Stateful inspection firewall.

    A **stateful inspection firewall** operates at the **transport layer** of the **OSI model**. This type of firewall monitors the state of **active connections** and inspects incoming and outgoing traffic based on the context of those connections. It can determine whether traffic is legitimate or not by examining packet headers and comparing them against established connection states.

25. What is the primary difference between an **Intrusion Detection System** (IDS) and an **Intrusion Prevention System** (IPS)?

    A. An IDS only detects attacks, while an IPS both detects and blocks attacks.

    B. An IDS only operates on the network layer, while an IPS operates on both the network and application layers.

    C. An IDS is passive and does not affect network traffic, while an IPS is active and can modify or block network traffic.

    D. An IDS only operates on inbound traffic, while an IPS also operates on outbound traffic.

    Answer: **C**. An IDS is passive and does not affect network traffic, while an IPS is active and can modify or block network traffic.

    An IDS is a security technology that monitors network traffic for signs of unauthorized activity or security policy violations. It is **passive** and does not modify or block network traffic. In contrast, an IPS is an **active** security technology that can modify or block network traffic based on predefined security policies.

26. Which of the following is an example of an anomaly-based detection method used by an IDS/IPS?

    A. Signature-based detection

    B. Protocol-based detection

    C. Statistical-based detection

    D. Behavioral-based detection

    Answer: **D**. Behavioral-based detection.

    Behavioral-based detection is an **anomaly-based detection** method used by an IDS/IPS. It involves analyzing patterns of network behavior over time and identifying deviations from normal behavior that could indicate an attack or security policy violation. Signature-based detection, protocol-based detection, and statistical-based detection are all examples of **rule-based detection** methods.

27. What is the primary purpose of a **Virtual Private Network (VPN)**?

    A. To provide secure remote access to a private network

    B. To encrypt all network traffic for privacy

    C. To enable fast and reliable network connectivity

    D. To prevent unauthorized access to a private network

    Answer: **A**. To provide secure remote access to a private network.

    A VPN is a secure tunnel that connects a remote user or site to a private network over the internet. The primary purpose of a VPN is to provide **secure remote access** to a private network, allowing authorized users to access network resources and data from any location, while ensuring the privacy and security of the transmitted data.

28. Which of the following VPN protocols is considered the most secure?

    A. PPTP

    B. L2TP

    C. SSTP

    D. IPsec

    Answer: **D**. IPsec.

    Internet Protocol Security (Ipsec) is considered the *most secure* VPN protocol because it provides strong encryption and authentication mechanisms to ensure the confidentiality, integrity, and authenticity of transmitted data. **Point-to-Point Tunneling Protocol (PPTP)** and **Layer 2 Tunneling Protocol (L2TP)** are older VPN protocols that are considered less secure due to vulnerabilities in their encryption algorithms. **Secure Socket Tunneling Protocol**

(SSTP) is a newer VPN protocol that uses SSL/TLS for encryption, but it is only supported on Windows operating systems.

29. Which of the following is an example of a cloud deployment model?

    A. Private cloud

    B. Public cloud

    C. Hybrid cloud

    D. All of these

Answer: **D**. All of these.

The three primary **cloud deployment models** are private cloud, public cloud, and hybrid cloud. Private cloud refers to a cloud environment that is dedicated to a single organization, while public cloud refers to a cloud environment that is shared by multiple organizations. Hybrid cloud refers to a combination of private and public cloud environments.

30. What is the shared responsibility model in cloud computing?

    A. A model that defines the roles and responsibilities of cloud service providers and customers in securing cloud environments

    B. A model that defines the level of access that customers have to cloud resources

    C. A model that defines the level of availability that cloud service providers guarantee to customers

    D. A model that defines the pricing structure for cloud services

Answer: **A**. A model that defines the roles and responsibilities of cloud service providers and customers in securing cloud environments.

The **shared responsibility model** is a model that defines the roles and responsibilities of cloud service providers and customers in securing cloud environments. Under this model, cloud service providers are responsible for securing the underlying cloud infrastructure, while customers are responsible for securing their own applications, data, and access to cloud resources.

31. What is encryption in the context of cloud security?

    A. A security control that prevents unauthorized access to cloud resources

    B. A security control that protects data by converting it into an unreadable format

    C. A security control that prevents denial of service attacks in cloud environments

    D. A security control that provides access control to cloud resources

Answer: **B**. A security control that protects data by converting it into an unreadable format.

Encryption is a **security control** that protects data by converting it into an unreadable format using a secret key. This helps to ensure the confidentiality and integrity of data stored in the cloud by making it unreadable to unauthorized users who may gain access to the data.

32. What is the primary goal of enterprise resiliency?

   A.   To prevent disruptions to business operations

   B.   To recover quickly from disruptions to business operations

   C.   To minimize the impact of disruptions to business operations

   D.   To increase the profitability of the organization

   Answer: **C**. To minimize the impact of disruptions to business operations.

   The primary goal of **enterprise resiliency** is to minimize the impact of disruptions to business operations. While preventing disruptions is important, it is not always possible, and organizations need to have plans in place to quickly recover from disruptions and minimize their impact.

33. Which of the following is *not* a component of enterprise resiliency?

   A.   Business continuity planning

   B.   Disaster recovery planning

   C.   Crisis management planning

   D.   Cost management planning

   Answer: **D**. Cost management planning.

   Cost management planning is not a component of **enterprise resiliency**. The three primary components of enterprise resiliency are business continuity planning, disaster recovery planning, and crisis management planning.

34. What is the difference between business continuity planning and disaster recovery planning?

   A.   Business continuity planning focuses on recovering from disruptions to business operations, while disaster recovery planning focuses on preventing disruptions from occurring.

   B.   Business continuity planning focuses on minimizing the impact of disruptions to business operations, while disaster recovery planning focuses on recovering from disruptions.

   C.   Business continuity planning focuses on the technology infrastructure required to support business operations, while disaster recovery planning focuses on the people and processes required to support business operations.

   D.   Business continuity planning and disaster recovery planning are the same thing.

   Answer: **B**. Business continuity planning focuses on minimizing the impact of disruptions to business operations, while disaster recovery planning focuses on recovering from disruptions.

   Business continuity planning and disaster recovery planning are both important components of **enterprise resiliency**, but they have different focuses. Business continuity planning focuses on *minimizing the impact* of disruptions to business operations by identifying critical business functions and processes, developing plans to ensure their continued operation during a disruption, and testing those plans regularly. Conversely, disaster recovery planning focuses on *recovering* from disruptions to IT systems and infrastructure by restoring data and systems as quickly as possible.

35. What is the purpose of data classification and labeling?

    A. To ensure that sensitive information is not disclosed to unauthorized parties

    B. To restrict access to data on a need-to-know basis

    C. To identify data that requires special handling based on its sensitivity and value

    D. All of these

    Answer: **D**. All of these.

    The purpose of data classification and labeling is to ensure that **sensitive information** is not disclosed to unauthorized parties, restrict access to data based on business need-to-know, and identify data that requires special handling based on its sensitivity and value.

36. Which of the following is *not* a common classification level for data?

    A. Top secret

    B. Sensitive but unclassified

    C. Restricted

    D. Public

    Answer: **A**. Top secret.

    While *top secret* is a common term used to describe highly sensitive government information, it is not a common classification level for data in the private sector. Common classification levels for data include public, sensitive but unclassified, confidential, and restricted.

37. What is data labeling?

    A. The process of classifying data based on its sensitivity and value

    B. The process of assigning metadata to data to provide additional context and meaning

    C. The process of encrypting data to ensure its confidentiality and integrity

    D. The process of removing data from a system or network to protect it from unauthorized access

    Answer: **B**. The process of assigning metadata to data to provide additional context and meaning.

    Data labeling is the process of assigning metadata to data to provide additional context and meaning. This can include information about the data's sensitivity level, how it should be handled and protected, and who is authorized to access it. Data labeling is an important component of **data classification**, as it helps to ensure that data is handled appropriately based on its sensitivity and value.

38. What is the primary objective of defining data retention?

    A. To ensure that data is accessible to authorized users whenever it is needed

    B. To ensure that data is backed up and recoverable in the event of a disaster

    C. To ensure that data is managed efficiently and cost-effectively throughout its life cycle

    D. To ensure that data is protected against unauthorized access and disclosure

    Answer: **C**. To ensure that data is managed efficiently and cost-effectively throughout its life cycle.

    This primary objective of defining **data retention** includes activities such as optimizing storage utilization, reducing storage costs, and ensuring that data is retained for an appropriate length of time.

39. What is the primary purpose of data life cycle management?

    A. To manage the creation, use, storage, and disposal of data throughout its life cycle

    B. To ensure that data is stored securely and access is restricted

    C. To ensure that data is backed up regularly and can be recovered in the event of a disaster

    D. To classify data based on its sensitivity and assign appropriate security controls

    Answer: **A**. To manage the creation, use, storage, and disposal of data throughout its life cycle.

    Data life cycle management refers to the process of managing the creation, use, storage, and disposal of data throughout its life cycle. This includes identifying data, classifying it based on its sensitivity, determining how long it should be retained, and establishing appropriate security controls for each stage of the data life cycle. The purpose of data life cycle management is to ensure that data is managed in a way that is consistent with business and regulatory requirements, and that it is protected against unauthorized access, loss, or theft.

40. What is the purpose of a data retention policy?

    A. To specify how data should be classified and labeled

    B. To ensure that data is stored securely and accessed only by authorized users

    C. To specify how long data should be retained and when it should be disposed of

    D. To ensure that data is backed up regularly and recoverable in the event of a disaster

    Answer: **C**. To specify how long data should be retained and when it should be disposed of.

    A **data retention policy** is a key component of data life cycle management and specifies how long data should be retained and when it should be disposed of. The policy should be based on business and regulatory requirements and should take into account factors such as the type of data, its sensitivity, and its value to an organization. The policy should also specify how data should be disposed of, such as through secure data destruction methods.

41. Which of the following is an example of a **multi-factor authentication (MFA)** mechanism?

   A. A username and password

   B. A fingerprint scan and a smart card

   C. A public key and a private key

   D. A PIN number and a security question

   Answer: **B**. A fingerprint scan and a smart card.

   MFA requires users to provide two or more forms of identification to access a system or resource. A fingerprint scan and a smart card are both examples of MFA mechanisms, as they require something the user has (the smart card) and something the user is (the fingerprint) to authenticate.

42. What is the purpose of access reviews?

   A. To determine the security controls needed for a resource

   B. To identify the risks associated with granting access to a resource

   C. To ensure that users have the appropriate level of access to a resource

   D. To audit the usage of a resource and detect any suspicious activity

   Answer: **C**. To ensure that users have the appropriate level of access to a resource.

   Access reviews are periodic evaluations of user access to resources, such as systems, applications, and data. The purpose of access reviews is to ensure that users have the appropriate level of access to a resource, based on their roles and responsibilities within an organization. Access reviews help to mitigate the risk of unauthorized access and reduce the likelihood of data breaches.

43. Which of the following is an example of a **segregation of duties (SoD)** control?

   A. An administrator has full access to a system and can perform all functions within that system.

   B. A developer has the ability to deploy code changes to production without any oversight.

   C. A quality assurance analyst reviews and approves code changes before they are deployed to production.

   D. A database administrator has the ability to modify user access controls and perform backups.

   Answer: **C**. A quality assurance analyst reviews and approves code changes before they are deployed to production.

   SOD controls ensure that no single person has the ability to perform all functions related to a particular business process. In the example given, the quality assurance analyst reviews and approves code changes, while the developer who made the changes is not allowed to deploy them to production without oversight. This helps to reduce the risk of errors or fraud and ensures that multiple people are involved in critical processes.

44. Which of the following is an example of symmetric encryption?

    A. RSA

    B. AES

    C. Diffie-Hellman

    D. ECC

    Answer: **B**. AES.

    The **Advanced Encryption Standard** (**AES**) is an example of symmetric encryption, which uses the same key to encrypt and decrypt data. **Rivest-Shamir-Adleman** (**RSA**), **Diffie-Hellman**, and **elliptic-curve cryptography** (**ECC**) are examples of asymmetric encryption, which uses different keys for encryption and decryption.

45. What is the purpose of a digital certificate in encryption?

    A. To encrypt data before it is transmitted

    B. To decrypt data after it is received

    C. To verify the identity of the recipient

    D. To verify the authenticity of the sender

    Answer: **D**. To verify the authenticity of the sender.

    A **digital certificate** is used in encryption to verify the authenticity of the sender, not to encrypt or decrypt data. A digital certificate is a type of digital signature that is issued by a trusted third party, such as a **certificate authority** (**CA**), and contains information about the sender's identity. When a recipient receives a message that has been digitally signed with a digital certificate, they can verify that the sender is who they claim to be and that the message has not been tampered with during transmission.

46. A company wants to implement encryption to protect sensitive data that is transmitted over the internet. Which type of encryption would be most appropriate for this scenario and why?

    A. Symmetric encryption

    B. Asymmetric encryption

    C. Hashing

    D. Digital signature

    Answer: **B**. Asymmetric encryption.

    Asymmetric encryption, also known as **public-key encryption**, uses different keys for encryption and decryption. This type of encryption is often used for secure communication over the internet, as it provides a *higher* level of security than **symmetric encryption**. With asymmetric encryption, a sender encrypts data using the recipient's public key, which can only be decrypted by the recipient's private key. This ensures that only the recipient can decrypt the data and provides a high level of confidentiality. Additionally, asymmetric encryption can also provide authentication and non-repudiation, which are important features for secure communication. Therefore, in this scenario, asymmetric encryption would be the most appropriate choice to protect sensitive data transmitted over the internet.

47. A company wants to ensure the integrity of the data stored in its database. Which of the following hashing algorithms would be most appropriate for this scenario and why?

    A. MD5

    B. SHA-1

    C. SHA-256

    D. HMAC

    Answer: **C**. SHA-256.

    SHA-256 is a *more secure* hashing algorithm than MD5 and SHA-1, which are vulnerable to **collision attacks**. SHA-256 produces a 256-bit hash value, which provides a high level of security and reduces the risk of collisions. Therefore, SHA-256 would be the most appropriate hashing algorithm for a company to ensure the integrity of the data stored in its database.

48. A company wants to securely store passwords in its database. Which of the following techniques would be most appropriate for this scenario?

    A. Encryption

    B. Hashing

    C. Salting

    D. Symmetric encryption

    Answer: **B**. Hashing.

    Hashing is a **one-way function** that converts plaintext into a fixed-length hash value. When storing passwords in a database, it is important to store only the hash value, not the plaintext password. This way, even if the database is compromised, the attacker will not have access to the actual passwords. When a user logs in, their plaintext password is hashed and compared to the stored hash value to authenticate them. Therefore, in this scenario, hashing would be the most appropriate technique to securely store passwords in a database. Salting, which involves adding random data to the plaintext password before hashing, can also be used to further increase the security of password storage.

49. A company wants to securely transmit sensitive information between its employees and clients over the internet. Which of the following PKI components would be most appropriate for this scenario?

    A.   A digital certificate

    B.   A CA

    C.   A public key

    D.   A private key

Answer: **A**. A digital certificate.

A **digital certificate** is a type of electronic credential that verifies the identity of the certificate holder and is used to establish a secure communication channel over the internet. In this scenario, a company should use digital certificates to authenticate the identity of both parties involved in the communication. This will ensure that only authorized parties can access the sensitive information being transmitted. The other components, such as the CA, public key, and private key, are necessary for the creation and management of digital certificates.

50. A company wants to ensure the confidentiality and integrity of its email communications. Which of the following PKI components would be most appropriate for this scenario and why?

    A.   A CA

    B.   A registration authority

    C.   A PKI policy

    D.   A digital certificate

Answer: **D**. A digital certificate.

Digital certificates are used to ensure the authenticity, confidentiality, and integrity of digital information, such as email communications. A digital certificate is a digitally signed document that binds a public key to an identity, such as a user or an organization. By verifying the digital certificate, the recipient can be sure that the email was sent by the claimed sender and that it was not tampered with in transit. Therefore, in this scenario, a digital certificate would be the most appropriate PKI component to ensure the confidentiality and integrity of email communications.

51. A company implements a PKI to secure its online transactions. Which of the following PKI components would be most important for this scenario?

    A.   A certificate revocation list

    B.   A CA

    C.   A certificate signing request

    D.   A PKI policy

Answer: **B**. A CA

A **Certificate Authority** (**CA**) is responsible for issuing digital certificates to users and organizations. In a PKI, the CA acts as a trusted third party that verifies the identity of the certificate requester, issuing a digital certificate that binds their public key to their identity. For online transactions, it is important to use a trusted CA to ensure that the digital certificates are authentic and that the transactions are secure. Therefore, in this scenario, a CA would be the most important PKI component to secure the company's online transactions.

52. A company plans to conduct security awareness training for its employees. Which of the following topics should be included in the training to promote a culture of security within the organization?

    A. Social engineering attacks

    B. Network protocols

    C. Server management

    D. Database design

    Answer: **A**. Social engineering attacks.

    Social engineering attacks are one of the most common ways in which attackers gain unauthorized access to sensitive information or systems. By educating employees about the different types of social engineering attacks, such as phishing, pretexting, and baiting, the company can help its employees identify and avoid these attacks, which will promote a culture of security within the organization.

53. A company conducts security awareness training for its employees. Which of the following best practices should be emphasized to prevent unauthorized access to sensitive information?

    A. Regularly updating software applications

    B. Using strong passwords

    C. Using personal devices for work-related tasks

    D. Sharing login credentials with coworkers

    Answer: **B**. Using strong passwords.

    Using strong passwords is one of the *most effective* ways to prevent unauthorized access to sensitive information. A strong password should be at least eight characters long and should include a mix of uppercase and lowercase letters, numbers, and symbols. By emphasizing the importance of using strong passwords and encouraging employees to use password managers to generate and store complex passwords, the company can significantly reduce the risk of a security breach. The other options, such as regularly updating software applications, not using personal devices for work-related tasks, and not sharing login credentials with coworkers, are also important security best practices, but they are not as effective as using strong passwords in preventing unauthorized access to sensitive information.

54. Which of the following statements is true about data security and data privacy?

    A. Data security and data privacy are the same thing.

    B. Data security refers to protecting data from unauthorized access, while data privacy refers to protecting personal information from being misused or mishandled.

    C. Data privacy refers to protecting data from unauthorized access, while data security refers to protecting personal information from being misused or mishandled.

    D. Data security and data privacy are not important for organizations to consider.

    Answer: **B**. Data security refers to protecting data from unauthorized access, while data privacy refers to protecting personal information from being misused or mishandled.

    Data security and **data privacy** are often used interchangeably, but they are *not* the same. Data security refers to the protection of data from unauthorized access, theft, corruption, or destruction. On the other hand, data privacy refers to the protection of personal information from being misused or mishandled. Personal information can include a person's name, address, social security number, credit card number, or medical records. While data security is important to protect personal information, it is only one part of data privacy.

55. A company collects customer data for marketing purposes. Which of the following measures should the company take to ensure that it complies with data privacy regulations?

    A. Use encryption to protect the data while it is stored.

    B. Obtain consent from the customers before collecting their data.

    C. Implement firewalls to prevent unauthorized access to the data.

    D. Train employees on how to handle sensitive data.

    Answer: **B**. Obtain consent from the customers before collecting their data.

    Data privacy regulations, such as the **General Data Protection Regulation** (**GDPR**) and the **California Consumer Privacy Act** (**CCPA**) require companies to obtain consent from individuals before collecting their personal information. This means that the company must inform the customers about the purpose of the data collection, what data is being collected, and how the data will be used. Additionally, the company should only collect the minimum amount of data necessary to achieve its purpose, and it should ensure that the data is accurate and up to date. While measures such as encryption, firewalls, and employee training are important for data security, obtaining consent from customers is critical to comply with data privacy regulations.

56. In an information security department, which of the following KPIs would be most appropriate to measure the effectiveness of an organization's vulnerability management program?

    A. The number of vulnerabilities found and remediated

    B. The number of security incidents reported

    C. The percentage of employees who have completed security awareness training

    D. The average time it takes to respond to a security incident

    Answer: **A**. The number of vulnerabilities found and remediated.

    The number of vulnerabilities found and remediated is a good KPI to measure the effectiveness of an organization's vulnerability management program. It indicates how successful a program is at identifying and addressing potential security risks.

    KPIs are used to measure the effectiveness of various aspects of an organization's operations. In this scenario, **vulnerability management** is the focus, and measuring the number of vulnerabilities found and remediated provides a clear metric to evaluate the effectiveness of the program.

57. A company has implemented a new security program to reduce the risk of data breaches. Which of the following KPIs would be most useful to evaluate the program's effectiveness over time?

    A. The number of employees who have completed security awareness training

    B. The number of security incidents reported

    C. The average time it takes to detect and respond to a security incident

    D. The percentage reduction in the number of data breaches

    Answer: **D**. The percentage reduction in the number of data breaches.

    The percentage reduction in the number of data breaches is the most useful KPI to evaluate the effectiveness of the security program over time. It provides a clear metric to measure the program's success in reducing the risk of data breaches.

    KPIs are used to measure the effectiveness of various aspects of an organization's operations. In this scenario, the focus is on evaluating the **effectiveness** of a new security program to reduce the risk of data breaches. While the other options may provide useful metrics to evaluate the program, measuring the percentage reduction in the number of data breaches is the most clear and direct way to determine the program's impact over time.

58. *Scenario*: A financial institution wants to test the effectiveness of its access control measures. The organization uses a biometric system for employee authentication, and the security team has identified this as a high-risk area for potential security breaches. The security team decides to perform a control test to evaluate the effectiveness of the biometric system.

    Question: Which of the following would be the most appropriate approach to test the effectiveness of the biometric system?

    A. Hire external consultants to conduct a penetration test.

    B. Observe and document the daily activities of system users.

    C. Conduct a technical assessment of the biometric system.

    D. Perform a vulnerability scan of the system.

    Answer: **C**. Conduct a technical assessment of the biometric system.

    A **technical assessment** would involve testing the system's functionality and security controls to determine whether it is effective in preventing unauthorized access. This type of testing is typically conducted by IT professionals who have expertise in the system being tested.

    Options *A*, *B*, and *D* are not effective approaches to test the effectiveness of access control measures.

    Penetration testing involves attempting to exploit vulnerabilities in a system to gain unauthorized access, but hiring external consultants to conduct a penetration test would *not* be effective in determining whether the biometric system functions as intended.

    Observing and documenting the daily activities of system users would *not* provide any insights into whether the biometric system works as intended or whether unauthorized access is prevented.

    Vulnerability scanning involves identifying vulnerabilities in a system, but performing a vulnerability scan of the system would *not* be effective in determining whether the biometric system functions as intended.

59. *Scenario*: A healthcare company has implemented a security control that requires employees to use a security token to access the company's network. The security team has identified the need to test the effectiveness of the security control to ensure that it functions as intended.

    Question: Which of the following would be the best approach to evaluate the effectiveness of the security control?

    A. Conduct a phishing simulation.

    B. Perform a vulnerability scan of the network.

    C. Conduct a technical assessment of the security token.

    D. Observe and document employee network usage.

Answer: **C.** Conduct a technical assessment of the security token.

A **technical assessment** involves testing the functionality and security controls of the security token to determine whether it is effective in preventing unauthorized access. This type of testing is typically conducted by IT professionals who have expertise in the system being tested.

Options A, B, and D are not effective approaches to evaluate the effectiveness of the security control.

A phishing simulation tests employees' susceptibility to phishing attacks, but conducting such a simulation would *not* provide any insights into whether the security token functions as intended.

Vulnerability scanning involves identifying vulnerabilities in a system, but performing a vulnerability scan of the network would *not* be effective in determining whether the security token functions as intended.

Observing and documenting employee network usage would *not* provide any insights into whether the security token functions as intended or whether unauthorized access is prevented.

60. During which phase of the SDLC should security requirements be identified and documented?

    A. The planning phase

    B. The design phase

    C. The development phase

    D. The testing phase

    Answer: **A.** The planning phase.

    Security requirements should be identified and documented during the **planning phase** of the SDLC. This is important to ensure that security controls are designed and implemented throughout the entire SDLC, rather than being added as an afterthought.

    Option B (design phase) could also be a plausible answer, as security controls are often designed during this phase. However, it is important to identify security requirements *before* the design phase so that security is integrated into the design of the system.

    Options C (development phase) and D (testing phase) are too late in the SDLC to identify and document security requirements.

61. Which of the following is the main objective of the testing phase in SDLC?

    A. To identify and fix defects in a system

    B. To develop the system according to user requirements

    C. To design the system architecture

    D. To create a project plan

Answer: **A**. To identify and fix defects in a system.

The **testing phase** in SDLC is an important phase where a developed system is tested thoroughly to identify and fix any defects before it is released to the production environment. The main objective of this phase is to ensure that the system functions as expected and meets the requirements of the users.

62. Which of the following is *NOT* an approach for threat modeling?

    A. STRIDE

    B. DREAD

    C. SAST

    D. PASTA

    Answer: **C**. SAST.

    STRIDE, DREAD, and PASTA are well-known approaches to threat modeling, while **Static Application Security Testing (SAST)** is not a threat modeling approach.

63. Which of the following is a key component of a threat landscape analysis?

    A. Risk assessment

    B. Asset classification

    C. Vulnerability scanning

    D. Threat intelligence

    Answer: **D**. Threat intelligence.

    A **threat landscape analysis** involves the evaluation of all potential threats to an organization. Threat intelligence is a key component of this analysis, as it provides information on the types of threats that an organization is likely to face, including their motives, capabilities, and attack methods.

64. Which of the following is a limitation of threat modeling?

    A. It is time-consuming and requires significant resources.

    B. It is not effective to identify advanced persistent threats.

    C. It primarily relies on assumptions.

    D. It is only effective to identify technical vulnerabilities, not human factors.

    Answer: **C**. It primarily relies on assumptions.

    Threat modeling is a valuable technique to identify potential threats and vulnerabilities in a system or application. However, it is not a perfect process and relies on assumptions, which may not always accurately reflect the reality of an organization's threat landscape. It is important to keep this limitation in mind when using threat modeling as part of a larger security strategy.

65. In which line of defense do the operational controls, risk management, and mitigation strategies lie?

    A. The first line of defense

    B. The second line of defense

    C. The third line of defense

    D. None of these

    Answer: **A**. The first line of defense.

    The first line of defense includes the day-to-day operational activities of an organization, including controls, risk management, and mitigation strategies.

66. Which of the following functions does not fall under the third line of defense in the three lines of defense model?

    A. An internal audit

    B. Compliance

    C. Enterprise risk management

    D. Business operations

    Answer: **D**. Business operations.

    The third line of defense typically includes independent functions such as an internal audit, compliance, and enterprise risk management, which oversee and evaluate the effectiveness of the first and second lines of defense. Business operations belong to the first line of defense.

67. Which line of defense is responsible for implementing policies, procedures, and standards to ensure compliance with laws, regulations, and internal policies?

    A. The first line of defense

    B. The second line of defense

    C. The third line of defense

    D. The fourth line of defense

    Answer: **B**. The second line of defense.

    The second line of defense is responsible for establishing and overseeing policies, procedures, and standards to ensure compliance with laws, regulations, and internal policies. This line of defense also provides guidance and support to the first line of defense. The second line of defense is also known as *risk monitoring and oversight*.

68. How does the alignment of security goals and objectives with an organization's overall strategy help in achieving its mission?

    A. It ensures that security is integrated into all business processes and decision-making.

    B. It helps in achieving compliance with legal and regulatory requirements.

    C. It ensures that security controls are implemented uniformly across the organization.

    D. It helps in reducing the overall cost of security operations.

    Answer: **A**. It ensures that security is integrated into all business processes and decision-making.

    Aligning security goals and objectives with an organization's overall strategy helps to ensure that security is integrated into all business processes and decision-making. This approach ensures that security is not viewed as a separate function but as an integral part of the organization's operations.

69. Which of the following is a key consideration for security professionals while aligning security objectives with an organization's strategy?

    A. Maximizing the budget for security operations

    B. Achieving compliance with all legal and regulatory requirements

    C. Ensuring that security goals do not conflict with the organization's mission

    D. Implementing the latest security technologies and tools

    Answer: **C**. Ensuring that security goals do not conflict with the organization's mission.

    When aligning security objectives with an organization's strategy, security professionals should ensure that security goals do not conflict with the organization's mission. This helps to ensure that security does not hinder the organization's operations but rather supports them. Maximizing the budget or implementing the latest tools and technologies may not always align with the organization's mission and may not necessarily lead to better security outcomes. Similarly, implementing security tools to comply with all legal and regulatory requirements may or may not align with the immediate business objectives.

70. A company's board of directors has set a risk appetite statement that states that they are willing to accept moderate risks to achieve higher returns. Which of the following actions is most likely to be taken by the organization's risk management team in light of this statement?

    A. They will adopt a more risk-averse approach to mitigate potential risks.

    B. They will develop risk mitigation strategies that align with the company's risk appetite.

    C. They will reject any business proposal that is considered too risky.

    D. They will completely ignore the risk appetite statement and proceed with their own risk management approach.

Answer: **B.** They will develop risk mitigation strategies that align with the company's risk appetite.

The **risk appetite statement** guides the risk management team on what level of risk an organization is willing to accept in pursuit of its goals. Based on this statement, the team will develop strategies that align with the company's risk appetite to manage risks in a way that supports the achievement of the company's objectives.

Option *A* is incorrect because adopting a more risk-averse approach would contradict the company's risk appetite statement. Option *C* is incorrect because the company's risk appetite does not necessarily mean that it will reject any business proposal that is considered too risky; instead, the risk management team will need to develop strategies that align with the risk appetite to manage risks associated with such proposals. Option *D* is incorrect because ignoring the risk appetite statement would be against the principles of effective risk management.

71. A company has a low-risk appetite and a high-risk tolerance level. What does this mean for the company's risk management approach?

   A. The company is willing to take high risks to achieve its objectives.

   B. The company is willing to take low risks to achieve its objectives.

   C. The company will mitigate all potential risks regardless of the impact on its objectives.

   D. The company will avoid all risks that exceed its risk appetite.

   Answer: **B.** The company is willing to take low risks to achieve its objectives.

   A **low-risk appetite** indicates that the company prefers to take fewer risks to achieve its objectives. However, a **high-risk tolerance** level indicates that the company is willing to accept some level of risk to achieve its objectives. This means that the company's risk management approach will focus on managing and mitigating low risks while accepting some degree of moderate risk to achieve its objectives.

   Option *A* is incorrect because a low-risk appetite means that the company is not willing to take high risks. Option *C* is incorrect because mitigating all potential risks regardless of their impact on objectives would not be an effective risk management approach, and it would not align with the company's risk tolerance level. Option *D* is incorrect because avoiding all risks that exceed the company's risk appetite would be too restrictive and could limit the company's ability to achieve its objectives.

72. What is the difference between risk appetite and risk tolerance in the context of an organization's risk management framework?

    A. Risk appetite is the amount of risk an organization is willing to accept, while risk tolerance is the level of risk an organization can tolerate.

    B. Risk tolerance is the amount of risk an organization is willing to accept, while risk appetite is the level of risk an organization can tolerate.

    C. Risk appetite and risk tolerance are interchangeable terms for the same concept.

    D. Risk appetite and risk tolerance do not apply to an organization's risk management framework.

    Answer: **A**. Risk appetite is the amount of risk an organization is willing to accept, while risk tolerance is the level of risk an organization can tolerate.

    Risk appetite and risk tolerance are two different concepts in an organization's risk management framework. **Risk appetite** refers to the amount of risk an organization is willing to take to achieve its goals and objectives, while **risk tolerance** refers to the level of risk an organization can tolerate before it takes action to reduce the risk.

    Option *B* is incorrect because it describes risk appetite as risk tolerance and vice versa. Option *C* is incorrect because risk appetite and risk tolerance are not interchangeable terms. Option *D* is incorrect because risk appetite and risk tolerance are important concepts in an organization's risk management framework.

73. What is the relationship between risk appetite and an organizational strategy?

    A. Risk appetite influences the development of an organization's strategy.

    B. An organizational strategy has no impact on risk appetite.

    C. Risk appetite is irrelevant to an organization's strategy.

    D. Risk appetite and organizational strategy are independent of each other.

    Answer: **A**. Risk appetite influences the development of an organization's strategy.

    An organization's **risk appetite** is an important factor in the development of its strategy. Risk appetite helps to define the level of risk an organization is willing to take on in pursuit of its goals and objectives. This level of risk can influence the development of the organization's strategy, as it may impact the types of risks and actions the organization is willing to take to mitigate those risks.

    Option *B* is incorrect because organizational strategy can have an impact on risk appetite. Option *C* is incorrect because risk appetite is relevant to an organization's strategy. Option *D* is incorrect because risk appetite and organizational strategy are interdependent.

74. Which of the following statements best describes risk appetite and risk tolerance in an organization?

    A. Risk appetite is the amount of risk an organization is willing to take on, while risk tolerance is the level of risk an organization can withstand.

    B. Risk appetite is the level of risk an organization can withstand, while risk tolerance is the amount of risk an organization is willing to take.

    C. Risk appetite and risk tolerance are the same thing and can be used interchangeably.

    D. Risk appetite and risk tolerance are not relevant to an organization's risk management strategy.

Answer: **A**. Risk appetite is the amount of risk an organization is willing to take on, while risk tolerance is the level of risk an organization can withstand.

Risk appetite represents the amount of risk an organization is willing to accept in order to achieve its goals, while **risk tolerance** is the level of risk an organization can withstand without experiencing unacceptable consequences.

Organizations should have a clear understanding of their risk appetite and risk tolerance in order to make informed decisions about risk management strategies. Risk appetite and risk tolerance are two different concepts that are often used interchangeably, but they represent different aspects of an organization's approach to risk management.

75. Which of the following is an example of risk tolerance?

    A. An organization accepts a high level of risk because it is necessary to achieve its business objectives.

    B. An organization invests in a security control to reduce the likelihood of a security breach.

    C. An organization avoids certain activities or processes that carry a high level of risk.

    D. An organization conducts a risk assessment to identify potential threats and vulnerabilities.

Answer: **A**. An organization accepts a high level of risk because it is necessary to achieve its business objectives.

Risk tolerance refers to the level of risk an organization is willing to accept in order to achieve its goals. In this scenario, the organization accepts a higher level of risk in order to achieve its business objectives, indicating a higher level of risk tolerance.

Risk tolerance is an important factor in risk management, as it helps organizations make decisions about the level of risk they are willing to accept in order to achieve their goals. This decision should be based on careful consideration of the potential impact of the risk, as well as the organization's ability to manage it.

76. A recent security assessment has identified a vulnerability in your organization's web application that could potentially allow unauthorized access to sensitive data. What should be the first step in conducting a root cause analysis?

   A.   Evaluate the potential impact of the vulnerability on the organization.

   B.   Determine the likelihood of the vulnerability being exploited.

   C.   Identify the source of the vulnerability.

   D.   Assess the effectiveness of existing controls in mitigating the vulnerability.

   Answer: **C**. Identify the source of the vulnerability.

   The first step in conducting a root cause analysis of a vulnerability is to identify the **source** of the vulnerability. This may involve reviewing the code for the application, analyzing network traffic, or conducting interviews with staff who are familiar with the application. Once the source of the vulnerability is identified, further analysis can be conducted to determine the potential impact and likelihood of exploitation, as well as the effectiveness of existing controls in mitigating the vulnerability.

   Option *A* is incorrect because evaluating the potential impact of the vulnerability is a subsequent step that comes after identifying the source of the vulnerability. Option *B* is incorrect because determining the likelihood of exploitation is also a subsequent step that comes after identifying the source of the vulnerability. Option *D* is incorrect because assessing the effectiveness of existing controls is also a subsequent step that comes after identifying the source of the vulnerability.

77. During a security audit, a control deficiency is identified in an organization's access management system that could potentially allow unauthorized access to sensitive data. What is the most appropriate course of action?

   A.   Develop a risk management plan to address the deficiency.

   B.   Accept the risk and continue to operate without mitigation.

   C.   Implement compensating controls to mitigate the deficiency.

   D.   Ignore the deficiency and hope that it is not exploited.

   Answer: **C**. Implement compensating controls to mitigate the deficiency.

   When a **control deficiency** is identified, the organization should take steps to mitigate the deficiency. In this case, implementing compensating controls would be the most appropriate course of action to mitigate the risk of unauthorized access to sensitive data. Developing a risk management plan may be necessary if the deficiency is complex or requires significant resources to address. Ignoring the deficiency is not a responsible course of action and could potentially lead to a security breach.

Option *B* is incorrect because accepting the risk without mitigation is not an appropriate course of action. Option *D* is incorrect because ignoring the deficiency is not a responsible course of action and could potentially lead to a security breach. Option *A* is not required unless the compensating controls are determined.

78. What is the purpose of a risk register in risk management?

    A. To track risks identified during risk assessments

    B. To mitigate risks after they occur

    C. To document risks for legal purposes

    D. To ignore risks that are deemed unlikely to occur

    Answer: **A**. To track risks identified during risk assessments.

    The purpose of a **risk register** is to track risks identified during risk assessments, including their potential impact, likelihood of occurrence, and risk response plans.

    A risk register is a tool used in risk management to track risks that were identified during a risk assessment. The risk register typically includes a list of identified risks, their potential impact on the organization, the likelihood of occurrence, and any planned risk response strategies. The risk register is an important tool to monitor and manage risks throughout the risk management process.

79. Which of the following is a common element in a risk register?

    A. Risk identification

    B. Risk acceptance

    C. Risk avoidance

    D. Risk transfer

    Answer: **A**. Risk identification.

    Risk identification is a common element in a **risk register**, as it is used to document the risks that were identified during a risk assessment.

    A risk register typically includes a list of identified risks, their potential impact on the organization, the likelihood of occurrence, and any planned risk response strategies. As stated in the answer to the previous question, the risk register is an important tool to monitor and manage risks throughout the risk management process.

80. Which of the following is a limitation of a risk register?

    A. It cannot be used to track risks that have not yet been identified.

    B. It does not provide a comprehensive view of all risks facing an organization.

    C. It is only useful to track risks in a specific project.

    D. It cannot be used to track risks that have been mitigated.

    Answer: **B**. It does not provide a comprehensive view of all risks facing an organization.

    One limitation of a risk register is that it does not provide a comprehensive view of all risks facing an organization. While a risk register is a useful tool to track risks identified during a risk assessment, it may not provide a comprehensive view of all risks facing an organization. Other risks, such as emerging risks or risks that have not yet been identified, may not be included in the risk register. Additionally, a risk register may be limited in scope, only covering risks associated with a specific project or initiative. Finally, a risk register may not be useful to track risks that have been mitigated, as those risks may no longer be relevant or require ongoing monitoring.

81. The IT department of a company has identified several security threats that could potentially harm the organization. What type of risk analysis methodology should the company use to prioritize these threats based on their likelihood and impact?

    A. Quantitative risk analysis

    B. Qualitative risk analysis

    C. The Delphi technique

    D. Monte Carlo simulation

    Answer: **B**. Qualitative risk analysis.

    This methodology uses **subjective judgments** to evaluate risks based on their likelihood and impact. It is used when there is limited data available or when the risks are too complex to model using quantitative methods.

    Option *A* is incorrect because quantitative risk analysis is a numerical approach that relies on data to assess risks. Option *C* is incorrect because the Delphi technique is a consensus-based approach that involves collecting opinions from experts. Option *D* is incorrect because Monte Carlo simulation is a statistical technique used to estimate the probability of different outcomes.

82. A company has decided to implement a new IT system that will process sensitive customer data. What type of risk analysis methodology should the company use to determine the security controls that are necessary to protect this data?

    A. Quantitative risk analysis

    B. Qualitative risk analysis

    C. Threat modeling

    D. The Delphi technique

    Answer: **C**. Threat modeling.

    Threat modeling involves identifying potential threats to an IT system and evaluating the likelihood and impact of those threats. It is often used to identify the security controls that are necessary to mitigate risks.

    Option *A* is incorrect because quantitative risk analysis focuses on measuring risks in financial terms. Option *B* is incorrect because qualitative risk analysis is a high-level approach that may not provide enough detail to identify specific threats. Option *D* is incorrect because the Delphi technique is not designed to identify specific risks or security controls.

83. A company considers a new business venture that involves working with a new partner. What type of risk analysis methodology should the company use to evaluate the potential risks associated with this partnership?

    A. SWOT analysis

    B. Monte Carlo simulation

    C. Risk assessment

    D. Root cause analysis

    Answer: **C**. Risk assessment.

    Risk assessment involves identifying potential risks, evaluating the likelihood and impact of those risks, and determining the appropriate risk response strategies.

    Option *A* is incorrect because SWOT analysis is a strategic planning tool that identifies strengths, weaknesses, opportunities, and threats. Option *B* is incorrect because Monte Carlo simulation is a statistical technique used to estimate the probability of different outcomes. Option *D* is incorrect because root cause analysis is used to identify the underlying causes of problems, not to evaluate risks.

84. A financial institution has identified a risk of fraudulent activities being carried out by external parties. Which of the following risk response options is best suited to mitigate this risk?

    A.   Acceptance

    B.   Transfer

    C.   Avoidance

    D.   Mitigation

    Answer: **D**. Mitigation.

    Mitigation is the most appropriate risk response option for this scenario, as it involves reducing the likelihood or impact of a risk through the implementation of controls. The financial institution can implement controls to mitigate the risk of fraudulent activities, such as enhancing identity verification procedures or implementing transaction monitoring tools.

85. A software development company has identified a risk of a data breach due to inadequate access controls. Which of the following risk response options is most appropriate for this risk?

    A.   Acceptance

    B.   Transfer

    C.   Avoidance

    D.   Mitigation

    Answer: **D**. Mitigation.

    The software development company can implement appropriate access controls such as authentication, authorization, and encryption to mitigate the risk of data breaches. **Mitigation** is the most appropriate risk response option for this scenario, as it involves implementing controls to reduce the likelihood or impact of a risk.

86. An organization has identified a risk of potential downtime due to power outages. Which of the following risk response options is most appropriate for this risk?

    A.   Acceptance

    B.   Transfer

    C.   Avoidance

    D.   Mitigation

    Answer: **B**. Transfer.

    The organization can **transfer** the risk to a third-party service provider by outsourcing its IT infrastructure or purchasing a business continuity and disaster recovery insurance policy. This is the most appropriate risk response option for this scenario, as it involves transferring the risk to another party who can better manage or absorb the risk.

87. An organization has implemented a new security control to mitigate a high-risk threat identified in the risk assessment. Which of the following monitoring techniques would be most effective in determining the effectiveness of the control over time?

    A. Annual assessments of the control

    B. Regular penetration testing

    C. Continuous monitoring of the control

    D. Ad hoc vulnerability scans

    Answer: **C**. Continuous monitoring of the control.

    Continuous monitoring would be most effective in determining the effectiveness of the control over time. This involves continuously monitoring the control and its associated risk over time to ensure that it remains effective and to identify any potential issues as they arise.

88. An organization performs a review of its security controls to ensure that they function effectively. Which of the following monitoring techniques would be most appropriate to identify potential control deficiencies?

    A. Internal audits

    B. Penetration testing

    C. Risk assessments

    D. Vulnerability scans

    Answer: **A**. Internal audits.

    Internal audits would be most appropriate to identify potential control deficiencies. An internal audit is a comprehensive review of an organization's controls, policies, and procedures, designed to ensure that they operate effectively and to identify areas for improvement.

89. An organization has implemented a new security control to mitigate a high-risk threat identified in the risk assessment. Which of the following monitoring techniques would be most effective in identifying any potential issues with the control shortly after implementation?

    A. Annual assessments of the control

    B. Regular penetration testing

    C. Continuous monitoring of the control

    D. Ad hoc vulnerability scans

Answer: **D**. Ad hoc vulnerability scans.

Ad hoc vulnerability scans would be most effective in identifying any potential issues with the control shortly after implementation. These scans are conducted on an as-needed basis, allowing the organization to quickly identify any potential issues with the control shortly after it has been implemented.

90. What is the purpose of change management in IT operations management?

A.   To ensure that changes are made to IT systems in a controlled and systematic way

B.   To minimize the impact of changes on users and systems

C.   To maintain a record of changes made to IT systems

D.   All of these

Answer: **D**. All of these.

The purpose of **change management** in IT operations management is to ensure that changes are made to IT systems in a controlled and systematic way, minimize the impact of changes on users and systems, and maintain a record of changes made to IT systems. All of the preceding factors are important for effective change management.

91. Which of the following best describes IT asset management?

A.   The process of identifying and tracking the hardware and software assets of an organization

B.   The process of monitoring and maintaining the performance of IT systems

C.   The process of managing the IT infrastructure of an organization

D.   The process of managing the risks associated with IT assets

Answer: **A**. The process of identifying and tracking the hardware and software assets of an organization.

IT asset management is the process of identifying and tracking the hardware and software assets of an organization. This includes keeping an inventory of assets, monitoring their performance, and ensuring they are used effectively. It is an important part of IT operations management, as it helps organizations to manage their IT resources effectively.

92. Which of the following is true of incident management in IT operations management?

A.   It is the process of restoring normal service as quickly as possible after an incident.

B.   It is the process of identifying and resolving the root cause of incidents.

C.   It is the process of monitoring IT systems for potential incidents.

D.   It is the process of prioritizing incidents based on their severity.

Answer: **A**. It is the process of restoring normal service as quickly as possible after an incident.

Incident management in IT operations management is the process of restoring normal service as quickly as possible after an incident. This includes identifying, prioritizing, and resolving incidents to minimize their impact on users and systems. While identifying and resolving the root cause of incidents is important, that falls under **problem management**, which is a systematic process to identify, analyze, and solve recurring incidents to prevent future disruptions in services.

93. What is the primary goal of business continuity management?

    A. To ensure the continued operation of critical business functions in the event of a disruption or disaster

    B. To prevent all possible risks from impacting business operations

    C. To minimize the cost of disruptions to an organization

    D. To comply with regulatory requirements

Answer: **A**. To ensure the continued operation of critical business functions in the event of a disruption or disaster.

The primary goal of **business continuity management** is to ensure that an organization can *continue* to operate critical business functions in the event of a disruption or disaster. This includes developing plans, policies, and procedures to respond to and recover from disruptions or disasters, as well as testing and maintaining those plans to ensure they are effective.

Option *B* is incorrect because it is not possible to prevent all possible risks from impacting business operations. Option *C* is also incorrect because minimizing the cost of disruptions is important but not the primary goal of business continuity management. Option *D* is incorrect because compliance with regulatory requirements is only one aspect of business continuity management.

94. Which of the following is an essential component of business continuity management?

    A. Developing a comprehensive risk management plan

    B. Conducting regular employee training and awareness programs

    C. Maintaining up-to-date inventories of critical resources and dependencies

    D. All of these

Answer: **D**. All of these.

Developing a comprehensive risk management plan, conducting regular employee training and awareness programs, and maintaining up-to-date inventories of critical resources and dependencies are all essential components of business continuity management. A comprehensive risk management plan helps identify potential risks and develop strategies to mitigate or manage them. Regular employee training and awareness programs help ensure that employees understand their roles and responsibilities in the event of a disruption or disaster. Maintaining up-to-date inventories of critical resources and dependencies helps ensure that an organization can quickly identify and address any potential gaps in its continuity plans.

Although options *A*, *B*, and *C* are important, they are not the sole essential component of business continuity management.

95. What is the purpose of disaster recovery management?

   A. To prevent disasters from occurring

   B. To minimize the impact of disasters on an organization

   C. To identify potential disasters before they occur

   D. To recover from disasters as quickly as possible

Answer: **D**. To recover from disasters as quickly as possible.

Disaster recovery management is a set of processes and procedures that an organization uses to recover from a disaster or disruptive event. The primary goal is to minimize the impact of the disaster on the organization's operations and return to normal as quickly as possible. While prevention and identification of potential disasters are important, the primary focus of disaster recovery management is on the recovery process.

96. What is the difference between disaster recovery and business continuity?

   A. There is no difference between the two terms.

   B. Disaster recovery focuses on IT systems, while business continuity focuses on an entire organization.

   C. Business continuity focuses on IT systems, while disaster recovery focuses on an entire organization.

   D. Disaster recovery focuses on preventing disasters, while business continuity focuses on recovering from disasters.

Answer: **B**. Disaster recovery focuses on IT systems, while business continuity focuses on an entire organization.

While both disaster recovery and business continuity are related to *recovering* from disruptive events, they focus on different aspects of an organization's operations. **Disaster recovery** primarily focuses on the recovery of IT systems, data, and technology infrastructure, while business continuity encompasses a broader set of processes and procedures to ensure the continued operation of the entire organization. **Business continuity planning** includes disaster recovery as one of its components and also includes planning for other aspects, such as employee safety, communication plans, and alternative work arrangements.

97. What is the role of the **Chief Information Security Officer (CISO)** in an organization's security structure?

    A. To oversee the development and implementation of security policies and procedures

    B. To manage an organization's security operations and incident response

    C. To ensure that all employees are trained on security awareness and best practices

    D. All of these

    Answer: **D**. All of these.

    The CISO is responsible for overseeing the development and implementation of security policies and procedures, managing an organization's security operations and incident response, and ensuring that all employees are trained on security awareness and best practices.

98. What is the role of the **Information Security (IS)** department in an organization's security structure?

    A. To implement and maintain security controls

    B. To perform vulnerability assessments and penetration testing

    C. To monitor a network for security breaches

    D. All of these

    Answer: **D**. All of these.

    The IS department is responsible for implementing and maintaining security controls, performing vulnerability assessments and penetration testing, and monitoring the network for security breaches.

99. What is the responsibility of all employees in an organization's security structure?

    A. To report security incidents and potential vulnerabilities

    B. To follow security policies and procedures

    C. To attend security awareness training

    D. All of these

    Answer: **D**. All of these.

    All employees have a responsibility to report security incidents and potential vulnerabilities, follow security policies and procedures, and attend security awareness training. Security is everyone's responsibility in an organization, not just the responsibility of the IT or security department.

100. What is the purpose of policies and standards in an organization's security framework?

    A.   To provide guidance and direction for security-related activities and decisions

    B.   To enforce strict rules and regulations on employees and stakeholders

    C.   To limit access to sensitive information and resources

    D.   All of these

Answer: **A**. To provide guidance and direction for security-related activities and decisions.

Policies and standards are essential components of an organization's **security framework** that provide a structure for decision-making, behavior, and actions. They define expectations, requirements, and constraints that must be followed to achieve a secure environment. While policies and standards may have specific rules and restrictions, their primary purpose is to provide guidance for individuals and the organization as a whole to adhere to, in order to achieve their security goals.

Options *B* and *C* are incorrect because policies and standards should not be seen as mechanisms to enforce strict rules or limit access to resources but, rather, to provide guidance and direction for security-related activities. Option *D* is incorrect because while policies and standards may include specific rules and regulations, their main purpose is to provide guidance and direction.

Index

A

abrupt changeover 124
acceptable use policy 41
access control policy 41
access management 179
  principles 179, 180
accreditation 172
Align, Plan, and Organize (APO) 10
Amazon Web Services (AWS) 114, 152
anomalies
  exceptions, managing 117
  findings, managing 117
  issues, managing 117
  responding 117
application-level gateway firewall 149
Appraisal Subcommittee (ASC) 58
artificial intelligence (AI) 169, 173
assets 42
asset valuation
  significance 44
asymmetric encryption 182
  private key 182
  public key 182
AWS Elastic Beanstalk 153

B

backup policy 41
Bayesian analysis 92
blockchain 169, 173
bottom-up risk assessment approach 89
bow tie analysis (BTA) 92
brainstorming/interview 92
breach 73
bring your own device (BYOD) 172, 173
  benefits 173
Build, Acquire, and Implement (BAI) 10
Business Associate Agreement (BAA) 115
business continuity (BC) 160
business continuity plan (BCP) 42, 98, 160
  exercise 51
business impact analysis (BIA) 97, 160
  concepts 98
  versus risk assessment 97, 98
business objectives 36, 54

C

California Privacy Rights Act (CPRA) 189
Capability Maturity Model (CMM) 143
  maturity levels 146

Capability Maturity Model
Integration (CMMI) 145
cause and consequence analysis 92
cause and effect analysis 92
Center for Information Security (CIS)
reference link 117
certificate authority (CA) 186
certificate revocation list (CRL) 186
certificates 186
X.509 certificate 186, 187
certification 172
change advisory board (CAB) 118
change management 118
change management policy 42
checklist 92
checklist-based risk assessment 93
CIA triad 178
availability 178
confidentiality 178
integrity 178
circuit-level gateway firewall 149
cloud computing 152
security considerations 153, 154
cloud computing deployment models 153
community cloud 153
hybrid cloud 153
private cloud 153
public cloud 153
cloud computing service models 152
Infrastructure as a Service (IaaS) 152
Platform as a Service (PaaS) 153
Software as a Service (SaaS) 153
CMM framework 145
maturity 145
reference link 146
COBIT 134
COBIT 2019 framework
publications 9, 10

Common Vulnerability Scoring
System (CVSS) 83
community cloud 153
compliance 5
computer networks 146, 147
confidentiality 178
least privilege principle 178
need-to-know principle 178
configuration checks 84
configuration management 117, 123
Continuing Professional
Education (CPE) 23
continuous monitoring 71
control
evaluating, best practices 126
selecting 123
testing 126
control assessments 134
internal information system (IS) audit 134
penetration testing 135
self-assessments 134
third-party assurance 135
vulnerability assessment 134
control design 123
control implementation 123
post-implementation reviews 125
control implementation techniques 124
abrupt changeover 124
parallel changeover 124
phased changeover 124
Control Objectives for Information and
Related Technology (COBIT)
and ITIL 11, 12
GRC, implementing with 9, 10
control owner 107, 126
corrective action plan (CAP) 94, 115
cost-benefit analysis 109
credential 180

CRISC exam
　additional resources 27, 28
　certification requirements 23
　domains 19
　job practice areas 19-22
　outline 18
　structure 22, 23
current risk 100
cybersecurity 7, 50, 51
**Cybersecurity and Infrastructure
　Security Agency (CISA) 84**
cybersecurity professionals
　GRC, significance 7-9

D

data classification 162
　access control 163
　business impact 163
　location 163
　regulatory requirements 162
　type 163
data classification policy 41
data labeling 162, 163
data leakage prevention (DLP) 163
data life cycle management 159, 163, 164
　archiving stage 164
　creation stage 163
　destruction stage 164
　sharing stage 163
　storage stage 163
　use stage 163
Data Loss Prevention (DLP) tool 71
data management
　versus data governance 164, 165
data privacy 143, 189
　principles 190, 191
Data Processing Addendum (DPA) 115

data security
　versus data privacy 191
Deliver, Service, and Support (DSS) 10
Delphi method 92
**Department of Defense Architecture
　Framework (DODAF) 145**
digital signatures 184, 186
　encryption 186
disaster planning flow 99
disaster recovery (DR) 160
　concepts 98
disaster recovery plan (DRP) 42
Domain Name System (DNS) 150
downstream third parties
　versus upstream third parties 116
DREAD model 82
dynamic analysis 84

E

**electronically stored Personal Health
　Information (ePHI) 58**
emerging technologies 172
　AI 173
　blockchain 173
　BYOD 172, 173
　Internet of Things 173
　quantum computing 174
encryption 181
　asymmetric encryption 182
　symmetric encryption 181
Enterprise Architecture (EA) 143, 144
　application architecture 144
　business architecture 144
　data architecture 144
　diagrammatic representation 144
　technology architecture 144

enterprise resiliency 159, 160
enterprise risk management (ERM) 36
ethics 60
European Union (EU) 189
Evaluate, Direct, and Monitor (EDM) 10
events 73
    correlating, with incidents 74
event tree analysis (ETA) 92
exception management 117

F

Factor Analysis of Information
    Risk (FAIR) 4, 92
fault tree analysis (FTA) 92
Federal Enterprise Architecture
    Framework (FEAF) 145
Federal Financial Institutions
    Examinations Council (FFIEC) 58
Federal Information Security
    Management Act (FISMA) 59
Federal Risk and Authorization
    Management Program (FedRAMP) 59
Financial Institutions Reform, Recovery and
    Enforcement Act of 1989 (FIRREA) 58
Financial Institutions Regulatory
    and Interest Rate Control
    Act of 1978 (FIRA) 58
firewalls 149
    application-level gateway firewall 149
    circuit-level gateway firewall 149
    next-generation firewall 149
    packet filtering firewall 149
    stateful inspection firewall 149
fundamental pillars, information security
    availability 178
    CIA triad 179

confidentiality 178
integrity 178

G

General Data Protection Regulation
    (GDPR) 59, 189
Gmail 153
Google App Engine 153
governance 4
    versus management 6, 7
governance of enterprise IT (GEIT)
    purpose 32
governance, risk, and compliance (GRC) 4
    for cybersecurity professionals 7-9
    implementing, with COBIT 9, 10
    relationship, between components 5, 6
Government Community Cloud (GCC) 153
GRC program, components
    monitoring and reporting 6
    sponsorship 6
    stewardship 6
guidelines, policy documentation 40

H

hashing 183
    SHA256 hashing 184
Health and Human Services (HHS) 58
Healthcare Insurance Portability and
    Accountability Act (HIPAA) 58
Health Information technology
    for Economic and Clinical
    Health Act (HITECH) 59
human reliability analysis (HRA) 93
hybrid cloud 153
hybrid risk assessment 90

I

IAAA
accountability 180
authentication 179
authorization 180
identification 179
incident response (IR) 180
incident response policy 42
incidents 73
correlating, with events 74
industry advisories 84
information assurance 7
information security 177
access management 179, 180
digital signatures 184-186
encryption 181
fundamental pillars 178
hashing 183
information security policy (ISP) 41
information security principles 143, 177
information technology 143
Infrastructure as a Service (IaaS) 152
inherent risk 100
Institute of Internal Auditors (IIA) 47
intellectual property terms 43
internal risk assessments 134
Internet of Things (IoT) 169, 173
Intrusion Detection System (IDS) 149
Intrusion Prevention System (IPS) 150
ISACA Code
of Professional Ethics 61, 62
ISACA exam questions
knowledge-based questions 24
scenario-based questions 24
ISACA mindset
developing 24-26
issues 73

IT governance 32
risk practitioners role 33
ITIL
and COBIT 11, 12
IT operations 143
IT risk 32, 68
terminologies 32
IT risk management 33
ethics, affecting 61
importance 15
laws 58-60
life cycle 34
IT risk management life cycle 68, 106
example 70, 71
risk and control monitoring 34, 69
risk assessment 34, 69
risk categorization 69
risk identification 34, 68
risk reporting 34, 69
risk response and mitigation 34, 69
IT risk strategy 35

K

key indicators 137
key control indicators (KCIs) 137
key performance indicators (KPIs) 137
key risk indicators (KRIs) 137
selecting 137
SMART metrics 137

L

least privilege principle 180
log aggregation and analysis 131, 132
log aggregation 133
log sources 132
logging and monitoring policy 42

M

management self-identified
    issues (MSIIs) 134

Markov analysis 93

maximum tolerable downtime
    (MTD) 99, 161

Monitor, Evaluate, and Assess (MEA) 10

Monte Carlo analysis 93

multi-factor authentication (MFA) 180

N

National Institute of Standards and
    Technology (NIST) 84, 134

National Institute of Standards and
    Technology (NIST) Cybersecurity
    Framework (CSF) 12, 37

    detect category 14

    identify category 12

    protect category 13

    recover category 14, 15

    respond category 14

National Vulnerability Database (NVD) 80

networking devices 148

    bridge 148

    functioning 148

    gateway 148

    repeater 148

    router 148

    switch 148

Network Time Protocol (NTP) 133

next-generation firewall 149

non-disclosure agreement (NDA) 114

non-repudiation 179

    example 179

O

Office 365 153

Office for Civil Rights (OCR) 58

Operationally Critical Threat, Asset,
    and Vulnerability Evaluation
    (OCTAVE) 83, 93

organizational asset 42

    asset valuation 44

    data 43

    intellectual property 43

    people 43

    technology 43

organizational culture 38, 39

organizational structure 36

    RACI 37

organization, steps for threats
    and threat mitigation

    proactive and implement controls 80

    vulnerabilities, prioritizing 80

OSI model 147

    application layer 147

    data link layer 147

    network layer 147

    physical layer 147

    presentation layer 147

    session layer 147

    transport layer 147

P

packet filtering firewall 149

parallel changeover 124

password policy 41

PCI DSS 180

penetration test 84
 black-box testing 135
 gray-box testing 135
 white-box testing 135
Personally Identifiable
  Information (PII) 162
phased changeover 124
physical security policy 41
Platform as a Service (PaaS) 153
policies 39
policy documentation 39-41
 essential policies 41
 exception management 42
 hierarchy 40
policy documentation, essential policies
 acceptable use policy 41
 access control policy 41
 backup policy 41
 business continuity plan (BCP) 42
 change management policy 42
 data classification policy 41
 incident response policy 42
 ISP 41
 logging and monitoring policy 42
 password policy 41
 physical security policy 41
 risk assessment policy 42
post-implementation reviews 125, 126
 questions 125, 126
principles, data privacy
 accountability 190
 accuracy 190
 confidentiality 190
 consent 190
 data breach notification 190
 data minimization 190
 data retention 190
 privacy by design 190

 security 190
 transparency 190
privacy 177
privacy principles 177
private cloud 153
private key encryption 182
proactive control 123
procedures 40
Process for Attack Simulation and
  Threat Analysis (PASTA) 82
Professional Ethics
 ISACA Code 61, 62
progressive testing 126
project risk 171
 external factors 171
 internal factors 171
Protected Health Information (PHI) 58
public cloud 153
public key cryptography 188
public key encryption 182, 186
public key infrastructure (PKI) 188
 use cases 188, 189

Q

qualitative risk analysis 90
quantitative risk analysis 90
quantum computing 169, 174

R

RACI 37
 roles 37
reactive control 123
recovery objectives 161, 162
recovery point objective (RPO) 98, 161
recovery time objective (RTO) 98, 161
regressive testing 126

release management 117
reliability 160
request for proposal (RFP) 114
residual risk 100
resiliency 159
    versus BCP 161
    versus recovery 159
return-on-investment thesis 110
risk 78, 80
    current risk 100
    inherent risk 100
    residual risk 100
    types 99
    versus IT risk 67
risk acceptance 54, 108
risk analysis 89
    qualitative risk analysis 90
    quantitative risk analysis 90
risk and control assessment 42
risk and control monitoring 69, 133
    control assessments 134, 135
risk and control reporting 135
    considerations 135
    primary formats 136
risk appetite 54
risk assessment 69-71, 84
    bottom-up risk assessment approach 89
    concepts 52
    frameworks 91
    methodologies 89, 90
    requirements 71, 72
    risk assessment policy 42
    techniques 92, 93
    versus business impact analysis (BIA) 97, 98
risk avoidance 109
risk capacity 52
    versus risk tolerance 53
risk categorization 69, 70

risk identification 68, 70
risk management 4, 36
    concepts 52
    ethical challenges 60
    relationship, between ethics and culture 60
    risk appetite 52
    risk capacity 52, 53
    risk profile 52, 53
    risk tolerance 52, 53
risk management, organizational culture
    compliant 39
    proactive 39
    reactive 38
    resilient 39
    vulnerable 38
risk manager 78
    responsibilities 37
risk mitigation 108
risk optimization 109
risk owner 107
risk practitioner 126
    role, in IT governance 33
risk register
    significance 94
risk reporting 69, 71
    dashboards 136
    executive summary 136
    heat maps 136
    scorecards 136
risk response 108
risk response and mitigation 69, 71
risk response and monitoring 105, 106
risk tolerance 52
    versus risk capacity 53
risk transfer 108
risk treatment 108
role-based access control (RBAC) 180

S

safeguards 123

Sarbanes-Oxley Act (SOX) 180

SDLC risk 171

Secure by Design, Secure by Default, Secure in Deployment and Communication (SD3+C) 81

Secure File Transfer Protocol (SFTP) 115

security awareness training 189

security considerations, cloud computing

compliance 154

data breaches 154

disaster recovery 154

identity and access management 154

Service-Level Agreements (SLAs) 154

shared responsibility 153

vendor lock-in 154

security development life cycle (SDL) 81

Security information and event management (SIEM) 133

Security Operations Center (SOC) 14, 132

segregation of duties (SoD) 134, 180

semiquantitative risk assessment 90

service-level agreements (SLAs) 114

SHA256 hashing 183

Sherwood Applied Business Security Architecture (SABSA) 145

single point of failure (SPOF) 189

SMART metrics 137

sneak circuit analysis (SCA) 93

Software as a Service (SaaS) 153

Spoofing, Tampering, Repudiation, Information Disclosure, Denial of Service, and Elevation of Privilege (STRIDE) 81

standard operating procedures (SOPs) 40

standards, policy documentation 39

stateful inspection firewall 149

static analysis 84

structured what-if technique (SWIFT) 93

symmetric encryption 181

system development life cycle (SDLC) 169

development/acquisition phase 170

disposal phase 171

implementation phase 170

initiation phase 170

operation/maintenance phase 170, 171

phases 169, 170

T

TCP/IP model 147

application layer 147

data link layer 147

network layer 147

physical layer 147

transport layer 147

techniques, risk assessment

Bayesian analysis 92

bow tie analysis (BTA) 92

brainstorming/interview 92

cause and consequence analysis 92

cause and effect analysis 92

checklists 92

Delphi method 92

ETA 92

Factor Analysis of Information Risk (FAIR) 92

FTA 92

human reliability analysis (HRA) 93

lotus blossom method of brainstorming 93

Markov analysis 93

Monte Carlo analysis 93

OCTAVE 93

sneak circuit analysis (SCA) 93

structured what-if technique (SWIFT) 93

The Open Group Architecture Framework (TOGAF) 145

third-party risk management (TPRM) 113-116

need for 113, 114

third-party security risk assessment 116

threat 32, 77-79

threat actor 32

threat intelligence (TI) logs 132

threat modeling 80, 81

importance 83

steps 81

threat modeling methods

DREAD 82

Operationally Critical Threat, Asset, and Vulnerability Evaluation (OCTAVE) methodology 83

Process for Attack Simulation and Threat Analysis (PASTA) 82

security development life cycle (SDL) 81

STRIDE 81

Trike 83

Visual, Agile, and Simple Threat (VAST) methodology 83

threat, vulnerability, and risk

relationship between 78

Three Lines of Defense (3LoD) model 47-51

audit 49, 50

operational management 49

responsibilities 49

risk and compliance functions 49

Trike 83

U

upstream third parties

versus downstream third parties 116

User Datagram Protocol (UDP) 147

V

vendor security risk assessment 116

virtualization 152

Virtual Private Network (VPN) 152

Visual, Agile, and Simple Threat (VAST) methodology 83

VPN server 152

vulnerabilities, identifying tools 84

configuration checks 84

dynamic analysis 84

industry advisories 84

penetration test 84

risk assessments 84

static analysis 84

vendor security feeds/bulletins 84

vulnerability assessment scans 84

zero-day findings 84

vulnerability 32, 78, 80, 134

vulnerability analysis 83

vulnerability management program (VMP) 85

W

white hat testing 135
wireless access point (WAP) 151
wireless networks 151
wireless router 151
WPA2 151

X

X.509 certificate 186

Z

Zachman Framework 145
zero-day findings 84

`Packtpub.com`

Subscribe to our online digital library for full access to over 7,000 books and videos, as well as industry leading tools to help you plan your personal development and advance your career. For more information, please visit our website.

Why subscribe?

- Spend less time learning and more time coding with practical eBooks and Videos from over 4,000 industry professionals

- Improve your learning with Skill Plans built especially for you

- Get a free eBook or video every month

- Fully searchable for easy access to vital information

- Copy and paste, print, and bookmark content

Did you know that Packt offers eBook versions of every book published, with PDF and ePub files available? You can upgrade to the eBook version at `packtpub.com` and as a print book customer, you are entitled to a discount on the eBook copy. Get in touch with us at `customercare@packtpub.com` for more details.

At `www.packtpub.com`, you can also read a collection of free technical articles, sign up for a range of free newsletters, and receive exclusive discounts and offers on Packt books and eBooks.

Other Books You May Enjoy

If you enjoyed this book, you may be interested in these other books by Packt:

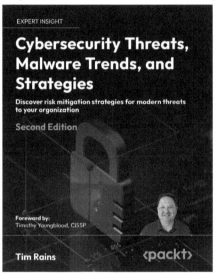

Cybersecurity Threats, Malware Trends, and Strategies - Second Edition

Tim Rains

ISBN: 978-1-80461-367-2

- Discover enterprise cybersecurity strategies and the ingredients critical to their success
- Improve vulnerability management by reducing risks and costs for your organization
- Mitigate internet-based threats such as drive-by download attacks and malware distribution sites
- Learn the roles that governments play in cybersecurity and how to mitigate government access to data
- Weigh the pros and cons of popular cybersecurity strategies such as Zero Trust, the Intrusion Kill Chain, and others
- Implement and then measure the outcome of a cybersecurity strategy
- Discover how the cloud can provide better security and compliance capabilities than on-premises IT environments

Digital Forensics and Incident Response - Third Edition

Gerard Johansen

ISBN: 978-1-80323-867-8

- Create and deploy an incident response capability within your own organization
- Perform proper evidence acquisition and handling
- Analyze the evidence collected and determine the root cause of a security incident
- Integrate digital forensic techniques and procedures into the overall incident response process
- Understand different techniques for threat hunting
- Write incident reports that document the key findings of your analysis
- Apply incident response practices to ransomware attacks
- Leverage cyber threat intelligence to augment digital forensics findings

Packt is searching for authors like you

If you're interested in becoming an author for Packt, please visit `authors.packtpub.com` and apply today. We have worked with thousands of developers and tech professionals, just like you, to help them share their insight with the global tech community. You can make a general application, apply for a specific hot topic that we are recruiting an author for, or submit your own idea.

Share Your Thoughts

Now you've finished *ISACA Certified in Risk and Information Systems Control (CRISC®) Exam Guide*, we'd love to hear your thoughts! Scan the QR code below to go straight to the Amazon review page for this book and share your feedback or leave a review on the site that you purchased it from.

`https://packt.link/r/1803236906`

Your review is important to us and the tech community and will help us make sure we're delivering excellent quality content.

Download a free PDF copy of this book

Thanks for purchasing this book!

Do you like to read on the go but are unable to carry your print books everywhere?

Is your eBook purchase not compatible with the device of your choice?

Don't worry, now with every Packt book you get a DRM-free PDF version of that book at no cost.

Read anywhere, any place, on any device. Search, copy, and paste code from your favorite technical books directly into your application.

The perks don't stop there, you can get exclusive access to discounts, newsletters, and great free content in your inbox daily.

Follow these simple steps to get the benefits:

1. Scan the QR code or visit the link below:

https://packt.link/free-ebook/9781803236902

2. Submit your proof of purchase
3. That's it! We'll send your free PDF and other benefits to your email directly